The Endocrine System

**Other titles in
Human Body Systems**

The Endocrine System

Stephanie Watson and
Kelli Miller

HUMAN BODY SYSTEMS
Michael Windelspecht, Series Editor

Greenwood Press
Westport, Connecticut • London

Library of Congress Cataloging-in-Publication Data

Watson, Stephanie, 1969–
    The endocrine system / Stephanie Watson and Kelli Miller.
        p.   cm.
    Includes bibliographical references and index.
    ISBN 0–313–32699–1 (alk. paper)
    1. Endocrine glands.   I. Miller, Kelli.   II. Title.
    QP187.W34   2004
    612.4—dc22   2004040447

British Library Cataloguing in Publication Data is available.

Copyright © 2004 by Greenwood Publishing Group, Inc.

Library of Congress Catalog Card Number: 2004040447
ISBN: 0–313–32699–1

First published in 2004

Greenwood Press, 88 Post Road West, Westport, CT 06881
An imprint of Greenwood Publishing Group, Inc.
www.greenwood.com

Printed in the United States of America

The paper used in this book complies with the
Permanent Paper Standard issued by the National
Information Standards Organization (Z39.48–1984).

10 9 8 7 6 5 4 3 2 1

Illustrations, unless otherwise credited, are by Sandy Windelspecht.

Contents

Color photos follow p. 62.

Series Foreword

Human Body Systems is a ten-volume series that explores the physiology, history, and diseases of the major organ systems of humans. An organ system is defined as a group of organs that physiologically function together to conduct an activity for the body. In this series we identify ten major functions. These are listed in Table F.1, along with the name of the organ system responsible for the activity. It is sometimes difficult to specifically define an organ system, because many of our organs have dual functions. For example, the liver interacts with both circulatory and digestive systems, the hypothalamus acts as a junction between the nervous and endocrine systems, and the pancreas has both digestive and endocrine secretions. This complex interaction of organs and tissues in the human body is still not completely understood.

This series is unique in that it provides a one-stop reference source for anyone with an interest in the human body. Whereas other references frequently cover one aspect of human biology, from anatomy and physiology to the prevention of diseases, this series takes a more holistic approach. Each volume not only includes a physiological description of how the system works from the cellular level upward, but also a historical summary of how research on the system has changed since the time of the ancients. This is an important aspect of the series and one that is frequently overlooked in modern textbooks. In order to understand the successes and problems of modern medicine, it is first important to recognize not only the achievements of the past but also the misunderstandings and challenges of the pioneers in medical research.

For example, a visit to any major educational institution reveals large lecture halls, where science instructors present material to the students on the

TABLE F.1. Organ Systems of the Human Body

Organ System	General Function	Examples
Circulatory	Movement of chemicals through the body	Heart
Digestive	Supply of nutrients to the body	Stomach, small intestine
Endocrine	Maintenance of internal environmental conditions	Thyroid
Lymphatic	Immune system, transport, return of fluids	Spleen
Muscular	Movement	Cardiac muscle, skeletal muscle
Nervous	Processing of incoming stimuli and coordination of activity	Brain, spinal cord
Reproductive	Production of offspring	Testes, ovaries
Respiratory	Gas exchange	Lungs
Skeletal	Support, storage of nutrients	Bones, ligaments
Urinary	Removal of waste products	Bladder, kidneys

anatomy and physiology of the human body. Sometimes these classes include laboratory sessions, but in the study of human biology, especially for students who are not bound for professional schools in medicine, the student's exposure to human biology typically centers on a two-dimensional graphic. Most educators accept this process as a necessary evil of the educational system, but few recognize that, in fact, the large lecture classroom is the product of a change in Egyptian religious beliefs before the start of the current era. During the decline of the Egyptian empire and the simultaneous rise of the ancient Greek culture, the Egyptian religious organizations began to forbid the dissection of the human body. This had a twofold influence on medicine. First, the ending of human dissections meant that medical professionals required lectures from educators, instead of participation in laboratory-based education, which led to the birth of the lecture hall. The second consequence would plague modern medicine for a thousand years. Stripped of their access to human cadavers, researchers studied other "lesser" animals and extrapolated their findings to humans. The practices of the ancient Greeks were passed on over the ages and became the basis for the study of modern medicine. These traditions continue to this day throughout the educational institutions of the world.

The history of human biology parallels the development of modern science. In the seventeenth century, William Harvey's study of blood circulation challenged the long-standing belief of the ancient Greeks that blood was produced in the liver and consumed in the tissues of the body. Harvey's pioneering experimental work had a strong influence on others, and within a century the legacy of the ancient Greeks had collapsed. In the eighteenth century a group of chemists who focused on the chemical reactions of the human body, called the iatrochemists, began to apply chemical laws to human physiology. They were joined by the iatrophysicists, who believed that the human body must operate under the physical laws of the universe. This in turn led to the beginnings of organic chemistry and biochemistry in the nineteenth century, as scientists focused on identifying the building blocks of living cells and the chemical reactions that they utilize in their metabolism.

In the past century, especially in the last three decades, the rapid advances in technology and scientific discovery have tended to separate most sciences from the general public. Yet despite an ongoing trend to leave the majority of the physical sciences to the scientists, interest in human biology has actually increased among the general population. This is primarily due to medical discoveries that increase not only lifespan but also healthspan, or the number of years that people live disease free. But another important aspect of this trend is the desire among the general public to be able to ask intelligent questions of their physicians and seek additional information on prescribed medications or procedures. In many cases this information serves as a system of checks and balances on the medical profession, ensuring that the patient is kept well informed and aware of the fundamentals regarding the procedure.

This is one of the most remarkable ages in the study of human biology. The recently announced completion of the Human Genome Project is an indication of how far biology has progressed. Barely fifty years ago, scientists were first discovering the structure of DNA. They now are in possession of an entire encyclopedia of human genetic information, and although they are not yet exactly sure what the content reveals, scarcely a week goes by without a researcher announcing a medical discovery that was made possible by the availability of the complete human genetic sequence. Coupled to this are the advances in the development of pharmaceuticals and treatments that were unheard of less than a decade ago.

But these benefits to society do not come without a cost. The terms stem cells, cloning, and gene therapy no longer belong to the realm of science fiction. They represent advances in the sciences that may hold the key to increased longevity. However, in many cases they also produce ethical and moral questions of society: Where do medical researchers obtain the embryonic stem cells for their work? Who will determine if humans can be cloned? What are the risks of transgenic organisms produced by gene therapy? These are just a few of the potential conflicts that face modern society. Only a well-educated general public can intelligently survey the pros and cons of an ethical or moral decision regarding medical science. Armed with information, concerned people can participate in the democratic process of informing their elected officials of their concerns. Science education is an important aspect of citizenship, and thus the need for series such as this to present information to the general public.

This volume covers the biology of the endocrine system. The endocrine system is the organ system of the body that is involved with the regulation and secretion of hormones. Hormones are chemicals that are secreted by one gland or organ of the body to influence the operation of a second, usually more distant, organ. The interactions of multiple organs via chemicals make the endocrine system one of the most complicated and least understood organ systems of the human body. While the nervous system provides us with almost instantaneous responses to stimuli, the endocrine system is responsible for ensuring that the internal conditions of the body remain stable for long periods of time. Blood sugar levels, tissue hydration, and calcium levels are all maintained under strict operational guidelines by the endocrine system. Problems with the endocrine system are often difficult to diagnose. Because most people are not familiar with the organs of this system, there is a definite need for a reliable reference that provides not only an overview of the endocrine system but also descriptions of some of the more common ailments and diseases that are associated with this system.

The ten volumes of the *Human Body Systems* are written by professional authors who specialize in the presentation of complex scientific ideas to the general public. Although any book on the human body must include the terminology and jargon of the profession, the authors of this series keep it to a minimum and strive to explain the concepts clearly and concisely. The series is ideal for the public libraries as well as for secondary school and introductory college libraries. In addition, medical professionals or anyone with an interest in human biology would find this series a useful addition to their personal library.

Michael Windelspecht
Blowing Rock, North Carolina

Acknowledgments

The authors would like to thank the Watson and Miller families for their support during the production of this book. They would also like to thank the National Institutes of Health and National Library of Medicine and to acknowledge the contributions of Michael Windelspecht, Debra Adams, Sandy Windelspecht, and Elizabeth Kincaid, without whom this book would not have been possible.

Introduction

To comprehend the role of the endocrine system and the pathways by which hormones affect biologic processes, it is first necessary to understand the chemical makeup of hormones, to learn how they are secreted and by what mechanisms they reach and interact with their target tissues, and to discern the complex relationship between the endocrine and nervous systems. These topics are covered in Chapter 1. Chapter 2 discusses the anatomy of the endocrine glands: the hypothalamus, pituitary, thyroid, parathyroids, adrenal, sex glands (testes and ovaries), and endocrine pancreas, as well as tissues throughout the body that are not considered classic endocrine organs but that secrete hormones or hormone-like substances. Subsequent chapters cover each endocrine gland in depth, devoting individual chapters to the hypothalamus and pituitary, the thyroid and parathyroids, the adrenal gland and endocrine pancreas, and the sex glands. Chapter 7 tracks the evolution of the endocrine system from fetal development to hormonal changes that occur late in life.

The next group of chapters chronicles the historic developments and major breakthroughs in endocrinology from ancient times to the new millennium. Chapter 8 details the earliest challenges and explores the long road to discovery over several centuries. Readers will learn about the advent of insulin therapy, the first therapeutic use of radioactive iodine, and other discoveries and inventions that shaped the field of endocrinology. The comprehensive chapter describes the work of Galen, the most important physician in ancient Greco-Roman times and one of the earliest pituitary researchers. It also tells the story of Albrecht von Haller, whose research regarding the ductless glands is considered a landmark in medical history. Chapter 9 peers into endocrinology's crystal ball, looking at the challenges and roadblocks that lie ahead. It also highlights the latest developments and breakthroughs in endocrine research.

The remaining chapters cover in detail the diseases and conditions affecting the thyroid, parathyroid, adrenal, and pituitary glands, as well as the endocrine pancreas. Each disease is accompanied by a description of risk factors, symptoms, diagnostic methodology, and treatment options.

This reference volume has been designed to provide a comprehensive overview of the endocrine system. It is geared to students, educators, patients, and anyone interested in the design and function of the endocrine glands and their respective hormones. High school and college students will find it useful for cross-referencing concepts and terminology from introductory biology and anatomy courses. Teachers can use it as a guide to support classroom curricula. And patients and their family members will find it an easy-to-use reference for researching a doctor's diagnosis or treatment options.

This work is designed to reach a general audience, and as such, every effort has been made to use simple, descriptive language. The reading level is appropriate for high school and introductory college students. When sophisticated terminology is necessary to describe a scientific or technical concept, it is highlighted in **bold** and a definition is provided in the glossary at the end of the book. Similarly, all acronyms are spelled out in an acronyms chapter. For readers seeking additional information on a particular condition, a list of organizations and Web sites has also been provided.

INTERESTING FACTS

▶ Ancient peoples called the pineal gland the "third eye" because they believed it held mystical powers. French philosopher René Descartes (1596–1650) thought that the pineal gland was the point at which the human soul met the physical body.

▶ The endocrine pancreas contains about 1,000,000 small endocrine glands called the islets of Langerhans.

▶ Special fluid sensors in the brain, called osmoreceptors, are so sensitive that they can detect a 1 percent fluctuation in the body's water concentration.

▶ The adrenal cortex produces more than sixty different steroid hormones, but only a handful are important to body function.

▶ Aldosterone, a steroid hormone produced in the adrenal cortex, acts upon the sweat glands to reduce the amount of sodium lost in the sweat. After a few days in a hot climate, sweat becomes virtually salt-free.

▶ During times of stress, the adrenal cortex can produce up to ten times the normal amount of cortisol.

▶ The fetal adrenal gland is larger than the adult gland in relation to body mass.

▶ In recent years, scientists have discovered that many industrial chemicals, pesticides, and heavy metals interfere with the endocrine systems of humans and wildlife by mimicking natural hormones.

▶ Endocrine disruptors have even been found in the breast milk of Inuit women in the remote Arctic, where known endocrine-disrupting chemicals are neither used nor produced.

▶ Scientists say that some deli wrap, food can linings, teething rings, vinyl toys, medical IV bags, and plastic bottles may seep small amounts of potential endocrine disruptors.

Fundamentals of the Endocrine System: Hormones and Their Actions

The intricate workings of the human body—its organs, muscles, bones, and tissues—form a sophisticated network, which is charged with the monumental task of sustaining life. Through extremes of cold, heat, hunger, thirst, and stress, the body must remain stable to avoid damage or even destruction. The body must also be adaptable, adjusting to each external and internal influence quickly and efficiently. To function properly in any environment, the body must maintain a constant internal state, with just the right balance of fluids and certain molecules. This equilibrium is called **homeostasis.**

For all of the various cells and tissues to work in harmony, they must communicate and coordinate with one another. A good analogy is a control tower at a busy airport. The control tower communicates and coordinates the movements of incoming and outgoing aircraft. Pilots rely on air traffic controllers to tell them when to take off, when to land, and which flight plans to follow. If suddenly all planes lost contact with air traffic control, the result would be chaos at the very least, and possibly even disaster.

Communication is every bit as essential in the human body. Without a network to integrate functions of the organs, muscles, nerves, and all other tissues, the body would virtually shut down. Ingested food would not be properly absorbed and utilized for energy, fluids and electrolytes would swing wildly up and down, and disease would easily set in, all with devastating effects.

To avoid these scenarios, the body has not one but two integrated command centers. These centers—the nervous system (which is covered in its own volume of this series) and the endocrine system—act as the body's control towers, sending out messages that coordinate the function of every cell. The nervous system sends out its messages via electrical impulses, which travel within nerve cells (neurons); and chemical signals (called **neurotransmitters**), which transmit those impulses across small gaps (called **synapses**) between the neurons. The endocrine system sends out its messages via chemical messengers called **hormones,** which travel through the bloodstream to act on cells in other parts of the body. Together, the two systems regulate every essential function from metabolism to growth and development.

The parts and functions of the endocrine and nervous systems are closely connected and synchronized. Nerves oversee the release and inhibition of endocrine system hormones, as well as blood flow to and from endocrine glands. Hormones, in turn, direct the nervous system by stimulating or inhibiting the release of neural impulses. The nervous system, as mentioned, also produces its own set of chemical messengers, called neurotransmitters, which enable nerve cells to communicate with one another. Collectively, the endocrine glands and endocrine-related parts of the nervous system are termed the **neuroendocrine system.** The study of this system is called **neuroendocrinology.**

The process by which the endocrine system coordinates bodily functions is complex, consisting of many interrelated parts and systems that oversee hormone production, secretion, and delivery. At the core of the system are the hormones—the messengers that transport endocrine commands throughout the body.

HORMONES

Hormones are the chemical signals by which the endocrine system coordinates and regulates functions such as growth, development, metabolism, and reproduction. The word "hormone" comes from the Greek word meaning "to set in motion." When a hormone is released into the bloodstream, it does just that: It sets in motion a chain of events that ultimately results in a desired reaction within cells that are receptive to its influence. The reaction is generally designed to either trigger or inhibit a physiological activity.

Hormones are produced by various tissues and secreted into the blood or **extracellular fluid.** According to the traditional definition, hormones travel through the bloodstream to work on tissues in distant parts of the body. But some hormones act locally without ever entering the bloodstream. They may exert their effects on cells close to where they were produced (called **paracrine** action), or they may act on the same cells that produced them (**autocrine** action).

Some hormones act upon just one type of cell; others influence many different cells. Similarly, some cells are receptive to only one hormone, while others respond to several hormones.

Hormones are not the only substances in the body that exert physiologic control on cells. A number of other chemical messengers act much like hormones. These include:

> *Neurotransmitters*: The nervous system, like the endocrine system, transmits messages to target tissues. But nervous system messages are made up of chemical signals called neurotransmitters. Unlike endocrine cells, which release their hormones into circulation, neurotransmitters are released into the gap where two **neurons** (cells in the brain that receive and transmit nerve impulses) meet (called a synapse). At the synapse, they bind to receptors on the receiving neuron. Some substances (such as **epinephrine, norepinephrine, dopamine, gastrin,** and **somatostatin**) serve double duty, acting as both hormones and neurotransmitters.
>
> *Growth factors*: Not to be confused with **growth hormones** produced by the endocrine system, **growth factors** are proteins that bind to receptors on the cell surface and stimulate or inhibit cellular division and proliferation. Some growth factors act on many different kinds of cells; others target one specific cell type. Examples include platelet-derived growth factor (PGDF), epidermal growth factor (EGF), transforming growth factors (TGFs), and **erythropoietin.**
>
> *Cytokines*: **Cytokines** are signaling peptides secreted by immune cells (as well as by other types of cells) in response to stress, allergic reaction, infection, or other potentially harmful stimuli. Much like endocrine hormones, cytokines either travel through the bloodstream or act locally on target cells. Once cytokines bind to their receptors, they trigger a biological effect within cells. Cytokines may influence cell growth, cell activation, or cell death (i.e., in the case of cancer cells). They also act directly upon the **hypothalamic-pituitary-target organ axis** of the endocrine system by increasing or decreasing hormone synthesis as part of the body's stress response. There are four major categories of cytokines: interleukins, interferons, colony stimulating factors, and tumor necrosis factors (TNF). (Cytokines are discussed further in the Lymphatic System volume of this series.)
>
> *Eicosanoids (fatty acid derivatives)*: These compounds are produced from **polyunsaturated fatty acids,** most commonly from the precursor arachidonic acid. Depending upon which enzymes act on arachidonic acid, it may be converted into one of several classes of hormone-like substances, including **prostaglandins,** prostacyclines, and thromboxanes. Although they are not technically hormones, **eicosanoids** act in much the same way to influence a variety of physiological processes, including smooth muscle contraction; kidney, immune system, and reproductive function; and calcium mobilization. Eicosanoids primarily exert a paracrine influence on nearby cells or an autocrine influence on the cells that produced them.

Just as a hormone does not always have to fit the classic definition, a hormone-producing tissue does not always need to reside within the endocrine system. Although hormones are primarily associated with the endocrine glands, tissues in other parts of the body (kidneys, liver, and heart, for example) can also produce and release them.

Hormone Modes of Action

Hormones are grouped according to their chemical structure (see Table 1.1). The structure of a hormone (i.e., whether it is water soluble or fat soluble) determines how it will travel through the bloodstream (alone or attached to a protein) and how it will bind to its receptor (fat-soluble hormones can travel through the membrane to receptors on the inside of the cell, while water-soluble hormones cannot pass through the membrane and must bind to receptors on the outside of the cell).

STEROID HORMONES

Steroid hormones (including **estrogen, testosterone,** and **cortisol**) are fat-soluble molecules produced from cholesterol. Because they generally repel water, steroid hormones travel through the blood attached to carrier proteins. Once they reach their target cell, steroid hormones pass through the cell membrane and bind to receptors in the cytoplasm and genes in the nucleus to regulate protein production.

AMINO ACID DERIVATIVES

Amino acid derivatives (such as epinephrine and norepinephrine) are water-soluble molecules derived from **amino acids** (compounds that form proteins). These hormones travel freely in the blood, but they cannot pass through the cell membrane, so they bind to receptors on the surface. Bind-

TABLE 1.1. Examples of Hormones within Each Class

Protein and peptide hormones	Antidiuretic hormone (ADH), follicle-stimulating hormone (FSH), glucagon, growth hormone (GH), insulin, luteinizing hormone (LH), oxytocin, prolactin (PRL), thyroid-stimulating hormone (TSH), thyrotropin-releasing hormone (TRH)
Steroid hormones	Aldosterone, cortisol, estrogen, testosterone
Amino acid derivatives	Epinephrine, norepinephrine, thyroxine, triiodothyronine

ing activates second messengers inside the cell that trigger enzymes or influence gene expression.

PROTEIN AND PEPTIDE HORMONES

Protein and peptide hormones (including **insulin, prolactin,** and growth hormone) are water-soluble hormones made up of amino acid chains. Like amino acid derivatives, peptide hormones circulate alone and bind to receptors on the cell surface.

Hormone Synthesis

PROTEIN AND PEPTIDE HORMONES

Protein and peptide hormones consist of chains of amino acids. These chains may number only a few amino acids in length, as is the case with many peptide hormones (**thyrotropin-releasing hormone [TRH]** contains only three amino acids); or they may contain more than 200 amino acids, as do many protein hormones (**follicle-stimulating hormone [FSH]** contains 204 amino acids).

A **glycoprotein** is a special type of protein hormone consisting of a protein connected to a glucose (sugar) molecule. Examples include **luteinizing hormone (LH),** follicle-stimulating hormone, and **thyroid-stimulating hormone (TSH).**

Peptide and protein hormones are produced (see Figure 1.1) in the endocrine cell under the direction of mRNA (messenger ribonucleic acid). The mRNA contains information that dictates the amino acid sequence of the protein. mRNA originates in the cell nucleus, then moves out into the cytoplasm, where it serves as a template for the amino acids to form an inactive molecule called a **preprohormone** (or **prohormone**). The prohormone is packaged into a secretory granule, which carries it to the cell surface. When the granule meets the cell membrane, an enzyme processes the prohormone to release the active hormone from the cell into circulation.

STEROID HORMONES

Steroid hormones are synthesized (see Figure 1.2) from cholesterol, about 80 percent of which comes from food and is transported through the blood plasma as **high-density lipoprotein (HDL)** particles. Included among the steroid hormones are the sex steroids (estrogen, **androgens** [testosterone], and **progesterone**) produced by the ovaries and testes, and the **glucocorticoids** (cortisol), **mineralocorticoids,** and androgens produced by the adrenal cortex.

To produce steroid hormones, enzymes convert cholesterol into a precursor molecule, called **pregnenolone,** in the cell mitochondria. Pregnenolone is then transported out of the mitochondria to the endoplasmic reticulum,

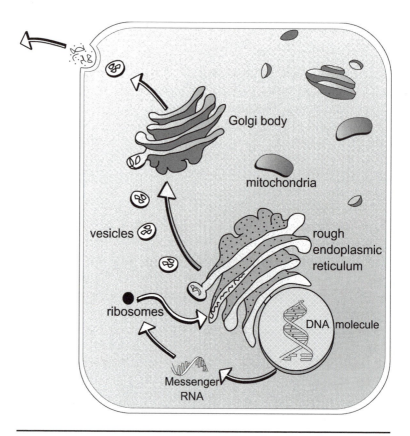

Figure 1.1. Protein hormone synthesis.

(1) mRNA directs the production of the preprohormone in the cytoplasm, (2) prohormone is packaged in a secretory granule, and (3) hormone is activated from the prohormone and released from the cell.

where enzymes break it down further to produce either another precursor(s) or the active steroid hormone. Unlike protein hormones, which require granules to transfer them to the cell surface, steroid hormones can make the trip on their own and exit the cell via diffusion across the membrane.

AMINO ACID DERIVATIVES

Unlike protein and peptide hormones, which consist of several linked amino acids, amino acid derivatives contain just one or two amino acids. Two major groups of hormones, both derived from the amino acid **tyrosine,** fall within this category: thyroid hormones (**thyroxine [T_4]** and **triiodothyronine [T_3]**) and **catecholamines** (epinephrine and norepinephrine, which are both hormones and neurotransmitters). Tyrosine reaches the endocrine cell via the bloodstream. Once inside the cell, enzymes transform the tyro-

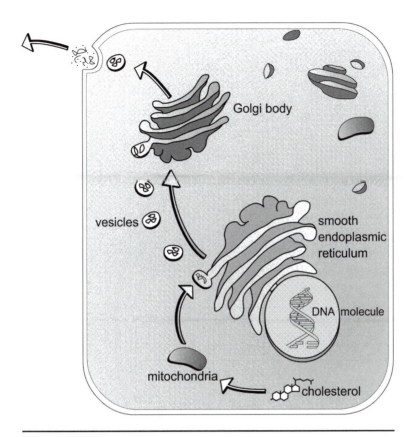

Golgi body

vesicles

smooth
endoplasmic
reticulum

DNA molecule

mitochondria

cholesterol

Figure 1.2. Steroid hormone synthesis.
(1) Enzymes convert cholesterol into pregnenolone in cell mitochondria, (2) pregnenolone is broken down into progesterone in the endoplasmic reticulum, and (3) progesterone is converted into the active steroid, which is released from the cell.

sine into the active hormone. In the case of thyroid hormones, iodine is
added to the modified tyrosine molecules. Amino acid hormones, like steroid hormones, can travel on their own across the cell membrane.

HORMONE TRANSPORT

Once a hormone is released from the cell, it travels through the bloodstream to the cell it will act upon. To get there, a hormone may either circulate alone (free) or bound to a carrier protein in the blood. As mentioned
above, amino acid, peptide, and protein hormones typically circulate free
because they are water soluble; steroid and thyroid hormones, which are fat
soluble, circulate bound to proteins. The advantage to binding is that the

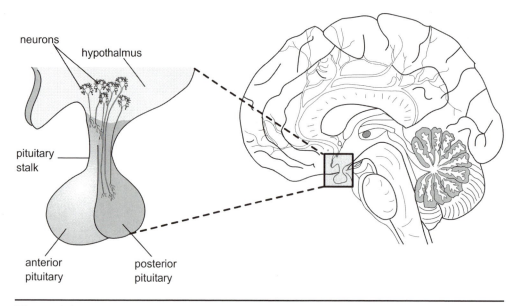

Figure 1.3. The hypothalamic-hypophyseal portal circulation.
Hormones released from the hypothalamus travel through blood vessels in the pituitary stalk to the anterior pituitary.

carrier protein helps the hormone navigate through all of the cellular traffic in the body to reach its target tissue. A bound hormone also stays longer in the blood than a free hormone, because its carrier protein holds it back from crossing a cell membrane. The level of bound and free proteins in the blood usually remains stable, because the proteins are in a concentration equilibrium, so each time a newly produced hormone is released into circulation, a hormone that is bound to a protein is freed.

Although most hormones travel through the bloodstream to reach their target cells, there are exceptions. One example is the **hypothalamic-hypophyseal portal system** (see Figure 1.3) through which releasing hormones secreted by nerve cells in the **hypothalamus** travel (via capillaries in the hypothalamus and veins in the pituitary stalk) directly to the **anterior pituitary** without ever entering the general circulation. See Chapter 3 for the interaction between the hypothalamus and pituitary glands.

TARGET CELLS AND RECEPTORS

A hormone cannot trigger a physiological reaction in just any cell—it is specifically designed to act upon *only* those cells that are receptive to it, which are called **target cells.** How does a hormone find its target cell? Each target cell comes complete with **receptors**—proteins that lie either on the surface of the cell or within the nucleus. The receptors exhibit specificity and bind only to the right hormone—or hormones—from a sea of other molecules. A cell's response depends upon the concentration of the hormone and the number of receptors to that hormone that it contains.

When the hormone binds to its receptor, it initiates a chain of events that ultimately alters the cell's function. The activated hormone-receptor complex can have one of three main effects: It can instruct cells to either make or stop making RNA from DNA (by the process of **gene transcription**), thus starting or stopping protein production; it can turn enzymes in the cell on or off, thus altering the cell's metabolism, or it can change the permeability of the cell membrane to allow in or shut out certain chemicals. If an individual lacks receptors for a particular hormone (or hormones), that hormone will not be able to do its job, and disease will often result.

Hormones not only trigger production of certain proteins within a cell; they may also block protein production or even block other hormones from binding to the cell receptor. Based on their effect, hormones are assigned to one of four classifications:

> *Agonists*: **Agonists** are hormones that bind to their receptor and elicit a specific biological response. For example, a glucocorticoid is an agonist for the receptor that binds cortisol.

> *Antagonists*: **Antagonists** are hormones that bind to the receptor but do not trigger a biological response. By occupying the receptor, the antagonist blocks an agonist from binding and thus prevents the triggering of the desired effect within the cell. For example, an antiandrogen is used to block the function of androgens in hormone therapy.

> *Partial agonist–partial antagonist*: A hormone that, when bound to the receptor, initiates a lesser biological response within the cell. By occupying the receptor, the partial agonist–partial antagonist blocks the potential action of an agonist, which could have generated a more significant biological response within the cell.

> *Mixed agonist-antagonist*: A hormone that exerts a different action on the receptor, acting as either an agonist or antagonist, depending upon the situation.

There are two types of receptors. Water-soluble hormones are unable to cross the membrane on their own because they are repelled by the fatty membrane that surrounds each cell, so they bind to receptors on the cell surface. Hormones that are fat soluble (such as steroids) are able to cross the membrane, so they bind to receptors inside the cell.

Cell Surface Receptors and Second Messenger Systems

Glucagon, catecholamines, **parathyroid hormone (PTH), adrenocorticotropic hormone (ACTH),** thyroid-stimulating hormone (TSH), and luteinizing hormone (LH) are water soluble and therefore cannot cross into

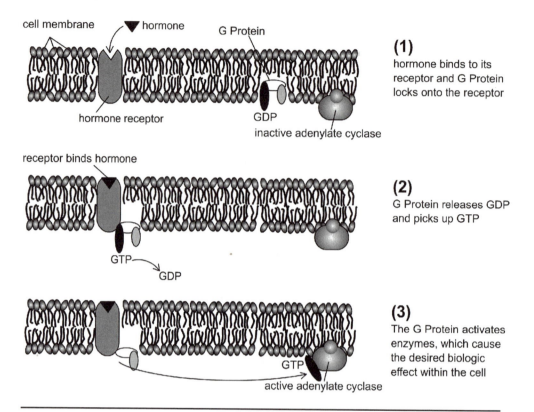

Figure 1.4. Second messenger system.

the cell. Instead, they bind to receptors on the surface of the cell membrane and trigger a cascade of events that leads to the desired biological response within the cell. For the message to pass from the hormone into the cell requires the efforts of second messengers (see Figure 1.4), which activate enzymes or other molecules inside the cell.

The hormone's actions are similar to those of a witness to a car accident. The witness acts as a first messenger, calling 911 and alerting the operator to the problem. The 911 operator is the receptor, taking the message and passing it along to the emergency crew, which acts as the second messenger, coming to the aid of the accident victims. (See "Other Second Messenger Systems" for further information.)

One of the primary second messenger systems involves G proteins. G proteins are like chemical "switches" that have to be triggered before the hormonal message can be passed into the cell. When a hormone binds to its receptor, the receptor changes shape and activates a G protein inside the cell. The G protein releases a guanine nucleotide it had been holding called

Other Second Messenger Systems

Growth factor receptors (receptors containing tyrosine kinase): Insulin, insulin-like growth factors (IGFs), and other growth factors bind to receptors containing the enzyme tyrosine kinase. Binding activates tyrosine kinase, which either directly or indirectly (through the production of second messengers) influences gene transcription.

Cytokine receptors (receptors associated with tyrosine kinase): Cytokines, erythropoietin (EPO), growth hormone (GH), oxytocin, and prolactin (PRL) bind to cytokine receptors. The receptor is not itself a tyrosine kinase, but the binding of the hormone with the receptor leads to a series of events that results in the activation of tyrosine kinase.

Calcium and/or phosphoinositides: Angiotensin II, antidiuretic hormone (ADH), epinephrine and norepinephrine, gonadotropin-releasing hormone (GnRH), and thyroid-releasing hormone (TRH) bind to calcium receptors. These receptors sense and respond to ions (such as calcium, magnesium, and sodium). Binding of the hormone and receptor opens or closes channels to let in or inhibit calcium and other ions.

guanine diphosphate (GDP), then grabs another, similar nucleotide called guanine triphosphate (GTP). Then the G protein goes to work, activating enzymes such as adenylyl cyclase. These enzymes produce a second messenger called cyclic adenosine monophosphate (cAMP). cAMP relays messages to effectors (molecules that regulate a series of chemical reactions) inside the cell, which lead to the desired biological reaction (for example, releasing glucose from cells when the body needs it for energy).

Intracellular Receptors

Receptors for steroid and thyroid hormones, as well as for vitamin D (a vitamin with hormone properties), are located inside the cell nucleus or cytoplasm. These hormones are fat soluble and can therefore cross the cell membrane on their own via simple diffusion. When they enter the cell, they meet up with and bind to their receptors, forming a hormone-receptor complex. The complex binds to parts of DNA in the cell nucleus called hormone response elements. Binding alters the DNA, resulting in the synthesis of a new protein.

HORMONE REGULATION

Hormones are so potent that just a tiny amount can exert powerful influences throughout the body. If too much or too little of a hormone is in circulation, the body can fall prey to serious disease (see Chapters 10 to 13). The effects of a particular hormone are related to its concentration in the bloodstream. Concentration is affected by the rate of production, the speed of distribution to target cells, and the speed at which the hormone is degraded after it is released from its receptor. All of these elements are strictly controlled by feedback loops or mechanisms, which measure and respond to changes within the body. Feedback loops ensure that enough hormones are produced to complete necessary tasks and keep the endocrine system tightly integrated with the nervous and immune systems.

Hormone Secretion

Some endocrine cells secrete their hormones at set times every day, every month, or even every year. Other cells secrete hormones following stimulation by other hormones, or in response to internal or external stimuli.

BIOLOGICAL RHYTHMS

Some hormones are released in regular patterns that follow a twenty-four-hour cycle (called circadian rhythms). Cortisol release, for example, rises in the early morning, gradually drops during the day, and stays very low during sleep. Other hormones follow a monthly, or even a seasonal, pattern. The pituitary gland, for example, releases luteinizing hormone and follicle-stimulating hormone in response to variations in a woman's monthly menstrual cycle.

HORMONAL INFLUENCE

The majority of hormones are regulated by other hormones (called **tropic hormones**), which either stimulate or inhibit their release based on the body's needs. The hypothalamus-pituitary control system is an example of tropic influence. The hypothalamus secretes several neurohormones, which signal the pituitary to release its own hormones. Pituitary hormones, in turn, direct the functions of several target organs.

INTERNAL AND EXTERNAL INFLUENCES

Many endocrine glands have their own mechanisms for sensing whether they need to release hormones. The endocrine-producing islet cells of the pancreas detect glucose levels in the blood, and release or inhibit insulin production as necessary. Sometimes, external factors are involved in hormone secretion. When a baby nurses from its mother's breast, the suckling action stimulates secretion of the hormones prolactin and **oxytocin.** Prolactin causes milk production in the mammary glands and maintains lactation. Oxytocin stimulates the release, or let-down, of milk into the nipple.

Feedback Mechanisms

Hormones regulate one another through feedback loops. The most basic feedback systems involve only one closed loop. More complex systems consist of a series of interrelated loops. Two main types of feedback systems exist:

Negative feedback: The most common type of feedback (see Figure 1.5) works much like a home air conditioning unit. When the temperature in the home rises to a preset level, a sensing mechanism turns on the air conditioner. After the air conditioning has run long enough to drop the temperature to a comfortable level, the shut-off mechanism is activated. Thanks to the feedback mechanism, the home is never allowed to get too cold or too warm.

In the body's negative feedback loop, a physiological change triggers the release of a particular hormone. Once the

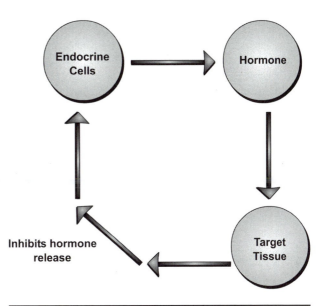

Figure 1.5. Negative feedback loop.
Rising blood levels of a hormone act upon the gland that secreted it to *inhibit* further production.

level of this hormone rises in the blood, it signals the endocrine cells that secreted it to stop producing it. Negative feedback prevents the overproduction of hormones, which could lead to disease.

An example of a simple negative feedback loop occurs after a person eats a piece of cake. After the cake is ingested, glucose (sugar) levels in the blood rise. In response to rising glucose levels, the endocrine cells of the pancreas release insulin. Insulin helps the cells take in and use glucose, lowering the amount of the sugar in the bloodstream. When blood glucose levels fall back to a normal level, insulin release is inhibited.

Positive feedback: As the name implies, a positive feedback loop (see Figure 1.6) stimulates, rather than inhibits, the production of a particular hormone. One example involves the release of oxytocin from the pituitary gland during childbirth. Oxytocin stimulates uterine contractions, which help push the baby out of the uterus. As levels of oxytocin in the blood rise, they trigger the pituitary to secrete even more of the hormone. Uterine contractions continue to increase until the child is finally born. Positive feedback is far less common than negative feedback, because it has the potential to contribute to dangerously high hormone levels.

More complex feedback systems involve several interrelated loops. One example is the hypothalamic-pituitary-target organ axis (see Figure 1.7), a multi-loop system that coordinates the efforts of the hypothalamus in the brain, the pituitary gland, and the target gland.

The hypothalamus, in response to reduced hormone levels in the blood-

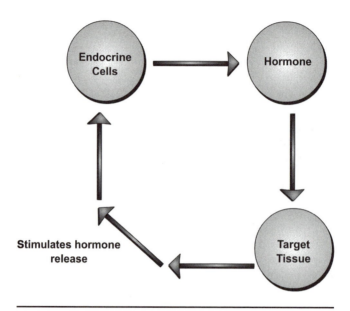

Figure 1.6. Positive feedback loop.
Rising levels of a hormone act upon the gland that secreted it to *stimulate* further production.

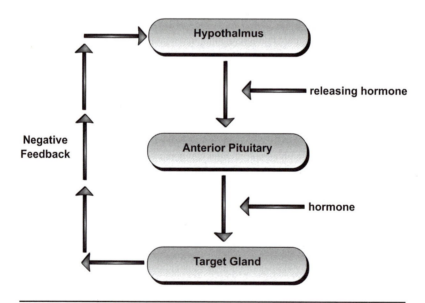

Figure 1.7. Hypothalamic-pituitary-target organ axis.
(1) Releasing hormone from the hypothalamus triggers pituitary hormone secretion, (2) pituitary hormone acts upon its target gland, and (3) as blood levels of pituitary hormone rise, the hypothalamus stops secreting its releasing hormone.

stream, stimulates pituitary hormone secretion. The pituitary hormone travels through the bloodstream to act on its target tissue(s). As the level of pituitary hormone in the blood rises, the hypothalamus stops secreting its releasing hormone. Consequently, the pituitary stops producing its own hormone, and blood levels of the hormone return to normal.

The hypothalamus has the ability to override the system, increasing or reducing hormone levels to adjust to physical and emotional stresses. Hormones from the target gland bind to nerve cells in the hypothalamus, which inhibit or trigger production of releasing hormones that influence pituitary hormone secretion. Without this mechanism, hormone levels would remain constant, even when they were needed in greater amounts to mediate a stress response.

Hormone Elimination

After hormones interact with their target cells and produce the desired result, they are no longer needed by the body. Most hormones are either converted to less active molecules or degraded by enzymes into an inactive form before being excreted in the urine or feces. Very few hormones are eliminated intact. Peptide hormones, catecholamines, and eicosanoids are all degraded by enzymes in the cell. Steroid hormones are metabolized into inactive forms and eliminated by the kidneys.

2

Endocrine Glands: Anatomy and Function

The endocrine system is made up of a complex network of glands (see Figure 2.1 and color insert), each of which secretes hormones that coordinate and regulate functions throughout the body. The pituitary, thyroid, parathyroids, adrenals, gonads (testes and ovaries), and endocrine pancreas are considered traditional endocrine glands with the primary purpose of secreting hormones into the bloodstream. But other, nonendocrine organs—including the heart, brain, kidneys, liver, skin, and gastrointestinal tract—can also secrete hormones.

By definition, endocrine glands release their hormones into the bloodstream to act upon target tissues elsewhere in the body. Endocrine glands should not be confused with **exocrine** glands, which secrete their substances to the outside of the body, to internal cavities such as the lumen of the intestines, or to other tissues through ducts (for example, the salivary and sweat glands).

The endocrine system is assigned several critical responsibilities, the most important of which are to maintain a constant internal environment (homeostasis); aid in growth, development, reproduction, and metabolism; and coordinate with the central nervous and immune systems.

WATER AND ELECTROLYTE BALANCE

For the body to function properly, it needs to maintain an internal balance of fluids and **electrolytes** (electrically charged chemical ions such as

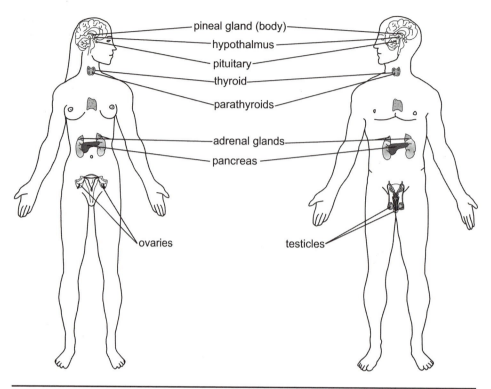

Figure 2.1. The endocrine system: female and male.

sodium, potassium, chloride, calcium, magnesium, and phosphate). More than 40 quarts (37 liters) of water circulate throughout the body. About two-thirds is **intracellular fluid,** located within the cells. About 75 percent of the remaining extracellular fluid is found in the tissue outside of the cells, and the other 25 percent is contained within the fluid portion of blood (plasma).

A rise in blood fluid volume (overhydration) can force the heart to work harder and dilute essential chemicals in the system. Too little water, or dehydration, can lead to low blood pressure, shock—and even death. The kidneys help to balance the fluid in the body by reabsorbing liquid into the bloodstream when levels get too low, or by eliminating excess fluids when levels rise too high. These processes occur under the direction of the endocrine system.

If the concentration of water drops too low (because not enough liquid was ingested or because fluid was lost through sweating, vomiting, or diarrhea), neurons called **osmoreceptors** send a message to the hypothalamus in the brain, which in turn tells the pituitary gland to secrete **antidiuretic hormone (ADH)** (also known as **vasopressin**) into the bloodstream. This hormone increases the permeability of the distal convoluted tubules and the collecting ducts in the nephrons of the kidneys, thus returning more fluid to the bloodstream. When more water is reabsorbed, the urine becomes more highly concentrated and is excreted in smaller volume. When the fluid concentration in the body is too high, ADH is not released. The distal convoluted tubules and collecting ducts are less permeable to water, and the kidneys filter out excess fluid, producing a larger volume of more dilute urine.

The kidneys must also maintain a balance of sodium, potassium, and other electrolytes in body fluids. To do this, they separate ions from the blood during filtration, returning what is needed to the bloodstream and sending any excess to the urine for excretion. Electrolyte levels are also directed by the endocrine system.

Sodium and potassium are two of the most important electrolytes, because without them, fluids would not be able to properly move between the intracellular and extracellular spaces. Sodium is the most abundant electrolyte in the extracellular fluid, and it also plays an important role in nerve and muscle function. The presence of too much sodium (a condition called hypernatremia) will send water from inside the cells into the extracellular region to restore balance, causing the cells to shrink. If nerve cells are affected, the result can be seizures and, in rare cases, coma. Too little sodium (called hyponatremia)—lost from excessive diarrhea, vomiting, or sweating—can send water into the cells, causing them to swell. Hyponatremia can lead to weakness, abdominal cramps, nausea, vomiting, or diarrhea. The swelling is even more dangerous if it occurs in the brain, where it can cause disorientation, convulsions, or coma.

Potassium assists in protein synthesis and is crucial for nerve and muscle function. Too little potassium can lead to a buildup of toxic substances in the cells that would normally pass into the extracellular fluid. To prevent a sodium-potassium imbalance, the cells use a mechanism called the sodium-potassium pump. This pump is a form of active transport (as opposed to the passive transport used in osmosis), which means that fluid can pass from one side of a semipermeable membrane to another, even if the concentration is already high on that side. But active transport requires energy to push molecules across the membrane. That energy is derived from adenosine triphosphate (ATP), a byproduct of cellular respiration. Once activated by ATP, the sodium-potassium pump pushes potassium ions into the cell while pumping sodium ions out of the cell until a balance is reached.

Endocrine hormones regulate the amount of sodium and potassium in the bloodstream. In the case of a sodium imbalance, an enzyme secreted by the kidneys, called **renin,** stimulates the production of the hormone **aldosterone** by the adrenal glands, which are located just above the kidneys. Aldosterone forces the distal convoluted tubules and collecting ducts in the kidneys to reabsorb more sodium into the blood. It also maintains potassium home-ostasis by stimulating the secretion of potassium by the distal convoluted tubule and collecting ducts when levels in the bloodstream get too high.

Parathyroid hormone (PTH), produced by the parathyroid glands, regu-lates levels of bone-building calcium and phosphate. When calcium con-centrations in the body drop, PTH pulls calcium from the bones, triggers the renal tubules to release more calcium into the bloodstream, and in-creases the absorption of dietary calcium from the small intestine. When too much calcium circulates in the blood, the thyroid gland produces an-other hormone, **calcitonin,** which causes bone cells to pull more calcium from the blood, and increases calcium excretion by the kidneys. PTH de-creases phosphate levels in the blood by inhibiting reabsorption in the kid-ney tubules, and calcitonin stimulates the bones to absorb more phosphate.

GROWTH, REPRODUCTION, AND DEVELOPMENT

Hormones play a role in virtually every aspect of human development, from fetal brain development (thyroid hormones) and sexual differentiation (luteinizing hormone [LH] and follicle-stimulating hormone [FSH]) to bone and cell growth (growth hormone) and reproduction (sex steroids and **go-nadocorticoids**).

Hormonal influence begins even before a child has been conceived. Hor-mones from the gonads (the ovaries and testes)—androgens, estrogens, and progestins—prepare the sperm and egg for conception. Once fertilization has occurred, maternal hormones establish and maintain the pregnancy, then cross the placenta to influence cellular functions in the fetus. The pla-centa itself produces many hormones that influence fetal development (**human chorionic gonadotropin [hCG]** and **placental lactogen**). Once the fetal endocrine glands develop, they take over from maternal and placental hormones and begin to produce their own hormones. Fetal sexual differen-tiation and development begin in the uterus and continue into puberty, all under the direction of hormones (testosterone, estrogen, and progesterone). Hormones play a role in childbirth by stimulating uterine contractions (oxy-tocin) and ripening the mother's cervix (prostaglandins). Once the child is born, hormones initiate milk production (prolactin) and lactation (oxytocin). Hormones even help a new mother nurture and bond with her new baby.

Metabolism

For the body to function properly, it needs energy. That energy comes from food, which the body converts into a usable form, stores, and then mo-bilizes to use when needed. Hormones carefully regulate these essential pro-cesses. Without insulin (from the endocrine pancreas), the body would starve, because it would be unable to turn glucose (sugar) from food into energy, and store that energy. Without glucagon (also from the pancreas), blood glucose levels would drop, and tissues would have no fuel to use during the periods between meals. Hormones not only influence how the body uses and stores glucose; they also oversee fat, protein, amino acid, and nucleic acid metabolism.

Bone

Hormones play an important role in bone growth, primarily through the mobilization of calcium. The bones hold 99 percent of the calcium in the human body. The tiny remainder is contained in the extracellular fluid. As indicated in the section on parathyroid hormone earlier in this chapter, hormones maintain a balance of calcium between the bones and extracellular fluid. Parathyroid hormone raises calcium concentration in the blood by enhancing reabsorption in the kidneys, increasing absorption of dietary calcium in the intestine, and mobilizing stored calcium from bone. Calcitonin has the opposite effect, lowering blood calcium levels by preventing calcium loss from bone and increasing excretion from the kidneys.

Central Nervous System

As mentioned in the first chapter, the endocrine and nervous systems work as a team. The nervous system, via the hypothalamus, releases or inhibits endocrine system hormones. Hormones, in turn, direct the nervous system by stimulating or inhibiting the release of neural impulses. Hormones can also indirectly influence the central nervous system through their effects on general metabolism. Neurons can be thought of as short-term regulators, while endocrine hormones are considered longer-term regulators.

Heart and Kidney Function

Hormones such as epinephrine and norepinephrine influence the cardiovascular system, affecting heart rate and blood pressure. Other hormones regulate fluid balance by altering the permeability of the ducts in the kidneys. Antidiuretic hormone (ADH) tells the kidneys to reabsorb more fluid into the blood. Aldosterone also regulates fluid concentration by increasing sodium reabsorption in the kidneys. Because water follows the sodium, less fluid is lost in the urine.

Immunity/Stress Response

Hormones are involved in the so-called fight-or-flight response, which helps the body cope with stress (i.e., injury, traumatic experience, or infection). Epinephrine and norepinephrine, the principal fight-or-flight hormones, increase heart and metabolic rates, constrict blood vessels, and raise the body's level of alertness. Hormones also play an important role in the immune response. Glucocorticoids, for example, suppress the immune system and have been used to treat a variety of autoimmune and inflammatory diseases.

ENDOCRINE GLANDS

The following section provides descriptions of the endocrine glands. Please refer to Table 2.1 for a list of the hormones produced by each gland.

Hypothalamus

The tiny, cone-shaped region at the base of the brain (see Figure 2.2) called the hypothalamus coordinates the neuroendocrine system, helps regulate metabolism, and controls the part of the nervous system that oversees a number of involuntary bodily functions (sleep, appetite, body temperature, hunger, and thirst). It also serves as the link between the nervous and endocrine systems.

The hypothalamus projects downward, ending at the pituitary stalk, which connects it to the pituitary gland. Together, the hypothalamus and pituitary (known collectively as the hypothalamic-pituitary axis [HPA]) direct the functions of the endocrine system. Although the pituitary has been termed the "master gland," the hypothalamus is the real control center behind the operation. The hypothalamus sends out messages (releasing or inhibiting hormones), which signal the pituitary to release—or stop releasing—its hormones. Pituitary hormones control the functions of virtually every endocrine gland in the body.

Pituitary

The pea-shaped pituitary gland (also known as the **hypophsis**) sits nestled in a cradle of bone at the base of the skull called the *sella turcica* ("Turkish saddle"). It is attached to the hypothalamus by the pituitary (hypophyseal) stalk, through which run the blood vessels and nerves (axons) that deliver hypothalamic hormones to the anterior pituitary.

As noted, the pituitary (see Figure 2.3) is often referred to as the "master gland" because it directs the functions of most other endocrine glands (including the adrenals, thyroid, and gonads [ovaries and testes]). In addition to stimulating other endocrine glands to release their hormones, the pituitary secretes several of its own hormones: growth hormone (GH), prolactin, and oxytocin (more on these in Chapter 3).

The pituitary gland is made up of three lobes: the anterior, intermediate, and posterior.

TABLE 2.1. Endocrine Glands and Their Hormones

Endocrine Gland	Hormone
Hypothalamus	Thyrotropin-releasing hormone (TRH) Gonadotropin-releasing hormone (GNRH) Corticotropin-releasing hormone (CRH) Growth hormone–releasing hormone (GHRH) Somatostatin (or growth hormone–inhibiting hormone [GHIH]) Prolactin-releasing hormone (PRH) Dopamine
Anterior pituitary	Thyroid-stimulating hormone (TSH) Growth hormone (GH) Gonadotropins (follicle-stimulating hormone [FSH], luteinizing hormone [LH]) Prolactin (PRL) Adrenocorticotropic hormone (ACTH)
Posterior pituitary	Antidiuretic hormone (ADH), or vasopressin Oxytocin
Pineal gland	Melatonin
Thyroid gland	Thyroxine (T_4) Triiodothyronine (T_3) Calcitonin
Parathyroid glands	Parathyroid hormone (PTH) Vitamin D
Adrenal cortex	Aldosterone (mineralocorticoid) Cortisol (glucocorticoid) Androgens and estrogen (gonadocorticoids)
Adrenal medulla	Epinephrine Norepinephrine
Endocrine pancreas	Insulin Glucagon Somatostatin Pancreatic polypeptide
Ovaries	Estrogens (estradiol, estriol, estrone) Progestogens (progesterone) Relaxin
Testes	Testosterone Dihydrotestosterone Estradiol

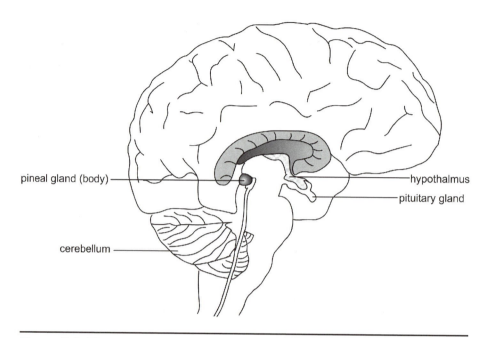

pineal gland (body)

hypothalmus

pituitary gland

cerebellum

Figure 2.2. The hypothalamus, pituitary, and pineal body.

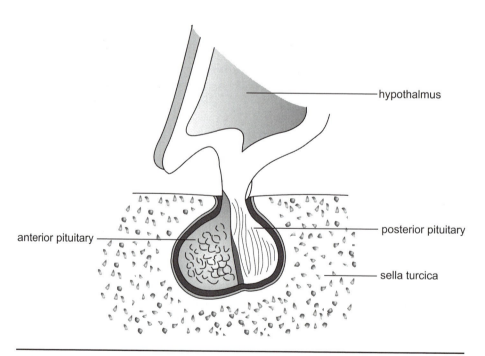

hypothalmus

anterior pituitary

posterior pituitary

sella turcica

Figure 2.3. The anterior and posterior pituitary gland.

Anterior pituitary: The anterior pituitary is composed of endocrine cells, which secrete hormones in response to stimulation by hypothalamic hormones. Anterior pituitary hormones, in turn, stimulate the adrenal glands (**adrenocorticotropic hormone [ACTH]**), thyroid gland (thyroid-stimulating hormone [TSH]), and ovaries and testes (follicle-stimulating hormone [FSH] and luteinizing hormone [LH]). The anterior pituitary also produces growth hormone (which stimulates growth of bone and muscle) and prolactin (which initiates milk production following childbirth).

Intermediate pituitary: The **intermediate pituitary** exists as a separate entity in animals, but only vestiges of this lobe remain in humans. Cells within the intermediate pituitary produce melanocyte-stimulating hormone, which controls skin pigmentation.

Posterior pituitary: Although the **posterior pituitary** is situated next to the anterior pituitary, it has very different functions. The posterior pituitary is an extension of the nervous system, made up primarily of axons and nerve endings that reach down from the hypothalamus. The posterior pituitary stores and releases hormones that are actually produced within the hypothalamus (see Chapter 3): antidiuretic hormone (ADH), which helps the body conserve water by increasing reabsorption in the kidney tubules; and oxytocin, which stimulates uterine contractions during childbirth and triggers the let-down of milk from the mother's breast when her infant nurses.

Pineal Gland

The small, cone-shaped pineal gland (see Figure 2.2) was once called the "third eye" and ascribed supernatural powers. It extends downward from the third ventricle of the brain, above and behind the pituitary gland. The pineal gland is composed of parts of neurons, but otherwise has no direct neural connection with the brain. Scientists know very little about the gland and what it does, but they do know that it secretes the hormone **melatonin,** which responds to light and dark, and communicates that information to the rest of the body. Melatonin influences circadian rhythms (the body's daily biological clock) and thus plays a role in functions regulated by night-day cycles, including reproduction and sleep/wake patterns.

Thyroid Gland

The largest endocrine gland in the body sits just below the larynx (voice box) and wraps around the trachea (windpipe). The thyroid gland (see Figure 2.4) resembles a butterfly, with its two lobes reaching out like wings on either side of a narrow strip of tissue called the isthmus. In a healthy adult, the thyroid weighs about 20 grams, but it can grow to several times this size (a condition called **goiter**; see Chapter 11).

The inside of the thyroid is composed of sacs called follicles. The follicles contain two types of cells: follicular cells and parafollicular cells. The majority of the cells are follicular, and it is in these cells that the thyroid

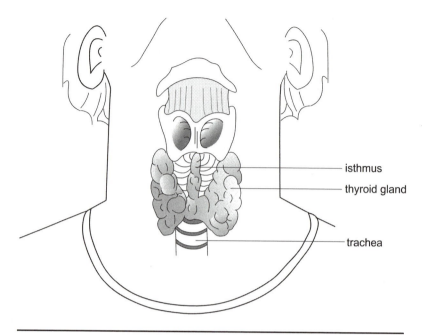

isthmus

thyroid gland

trachea

Figure 2.4. The thyroid gland.

hormones thyroxine (T_4) and triiodothyronine (T_3) are synthesized. In the spaces between the follicular cells are the parafollicular cells, which secrete the hormone calcitonin. The thyroid needs iodine to make these hormones.

Thyroid hormones play an important role in metabolism and calcium balance. Thyroxine and triiodothyronine are involved in energy regulation, heat mechanisms, protein synthesis, and cellular oxygen use. These activities are necessary for growth, development, reproduction, and the maintenance of a constant body temperature. Calcitonin preserves calcium in the body by increasing absorption of ingested calcium in the intestines, releasing calcium from the bones into the bloodstream when it is needed, and reabsorbing calcium into the bloodstream in the kidneys.

Parathyroid Glands

Most healthy adults have two pairs of oval-shaped parathyroid glands (see Figure 2.5), which lie next to the thyroid gland in the neck (the word "parathyroid" means "beside the thyroid"). In some instances, individuals may have fewer than or more than four parathyroid glands. Inside the glands are clusters of epithelial cells that produce and secrete parathyroid hormone, which is the most significant regulator of calcium levels in the blood. Calcium is essential for cell function as well as for bone formation.

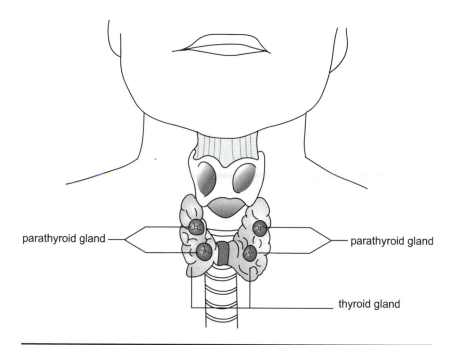

parathyroid gland

parathyroid gland

thyroid gland

Figure 2.5. The parathyroid glands.

Even small fluctuations in calcium levels can lead to muscle and nerve impairment.

Adrenal Glands

The adrenal glands (see Figure 2.6A and 2.6B) are often referred to as the fight-or-flight glands because they secrete hormones involved in the body's stress response. The small, triangular glands sit on top of each kidney, surrounded by a capsule of connective tissue. Each gland has access to a rich blood supply, through which it receives hormones from the pituitary and delivers its own hormones elsewhere in the body.

Like the pituitary gland, the adrenals are essentially two glands in one: an outer cortex and an inner medulla. Each region originates from a separate embryological source, functions separately, and produces its own distinct hormones.

MEDULLA

The inner portion of the adrenal takes up only about 10 percent of the gland, but it produces two crucial hormones: epinephrine (adrenaline) and norepinephrine (noradrenaline), referred to collectively as the catecholamines. These hormones (which are also neurotransmitters) maintain homeostasis

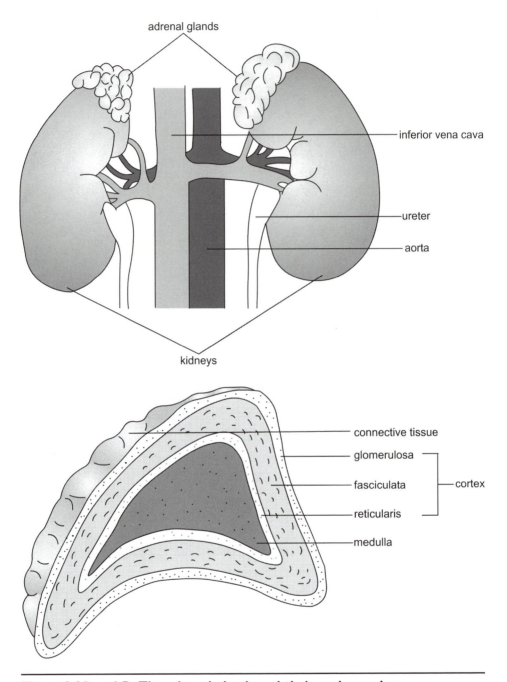

adrenal glands

inferior vena cava

ureter

aorta

kidneys

connective tissue

glomerulosa

fasciculata — cortex

reticularis

medulla

Figure 2.6A and B. The adrenal glands and their major regions.

when the body is confronted by emotional or physical stress. The medulla is an extension of the sympathetic nervous system and contains portions of nerve fibers. It is filled with chromaffin cells, inside of which are large numbers of vesicles that hold the catecholamines.

CORTEX

The dark outer region makes up the majority (about 90 percent) of the adrenal gland. The cortex produces and secretes the steroid hormones (glucocorticoids and mineralocorticoids), which influence metabolism, blood pressure, and sodium and potassium balance. Within the adrenal cortex lie three distinct zones:

> *Zona glomerulosa*: The outer zone secretes mineralocorticoids (chiefly aldosterone), which regulate water and sodium balance by increasing sodium reabsorption in the distal tubules of the kidneys and increasing potassium excretion in the urine. As more sodium is reabsorbed into the bloodstream, more water is also reabsorbed.

> *Zona fasciculata*: The middle zone, which makes up the bulk of the cortex (about 75 percent), produces the glucocorticoids (cortisol). Glucocorticoids maintain blood sugar levels and are also involved in metabolism and the immune response.

> *Zona reticularis*: The innermost zone is the thinnest layer of the cortex. In this zone, the gonadocorticoids (the sex hormones: androgens and estrogen) are produced. Gonadocorticoids influence the development of female and male sex characteristics.

Pancreas

The pancreas (see Figure 2.7) has two roles: It functions both as an endocrine and as an exocrine organ. As an exocrine organ, the pancreas releases digestive enzymes via a small duct into the small intestine. These enzymes break down carbohydrates, fats, and proteins from food that has been partially digested by the stomach. The exocrine pancreas also releases a bicarbonate to neutralize stomach acid in the duodenum (first portion) of the small intestine (more on these functions can be found in the Digestive System volume of this series).

In its role as an endocrine organ, the pancreas secretes the hormones insulin and glucagon, which help the body use and store its primary source of energy—glucose (sugar). The endocrine pancreas also secretes somatostatin, which is a primary regulator of insulin and glucagon release.

The pancreas is long and soft, and stretches from the duodenum of the small intestine almost to the spleen. It is divided into a head (its widest point), neck, body, and tail. The endocrine pancreas is made up of clusters

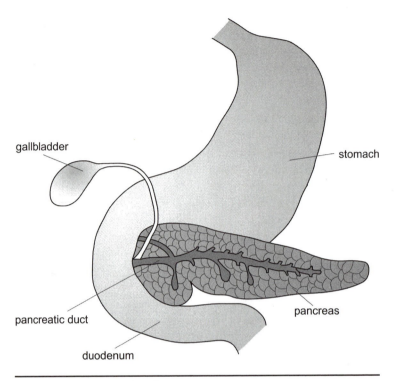

Figure 2.7. The endocrine pancreas.

of cells called the **islets of Langerhans,** in which the hormones insulin and glucagon are produced. There are about 1 million islets, but they make up only about 1 percent of the endocrine pancreas' total volume. The islets also contain parasympathetic and sympathetic neurons, which influence insulin and glucagon secretion.

Four different cell types exist within the islets (see Figure 2.8), each of which synthesizes its own hormones:

Alpha cells (A cells): Glucagon

Beta cells (B cells): Insulin

Delta cells (D cells): Somatostatin and gastrin

F cells: Pancreatic polypeptide

Each islet is supplied with numerous blood vessels, which carry pancreatic hormones elsewhere in the body. Although the islets make up just a tiny percentage of the pancreatic mass, they receive 10 to 15 percent of the entire blood flow that reaches the organ.

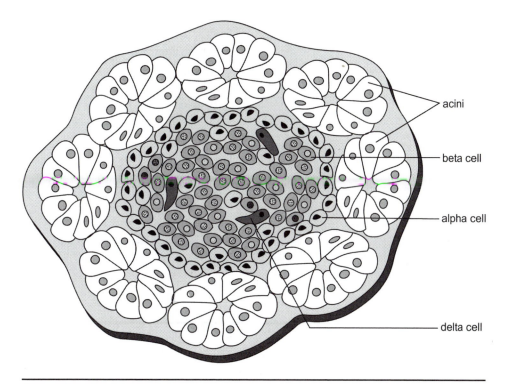

acini

beta cell

alpha cell

delta cell

Figure 2.8. Islets of Langerhans.

Sex Glands

The sex glands (ovaries in the female and testes in the male) serve as both reproductive and endocrine organs. They produce the eggs and sperm that form the basis of human life. They also synthesize and secrete the sex steroids—testosterone, estrogen, and progesterone. These hormones give males and females their individual sexual characteristics, and play a key role in reproduction.

The two bean-shaped ovaries (see Figure 2.9) sit on either side of the female uterus, just below the openings to the fallopian tubes. Like the adrenal glands, the ovaries contain an outer cortex and an inner medulla. The medulla consists primarily of connective tissue containing blood vessels, smooth muscle, and nerves. The real activity occurs in the larger outer cortex, which holds the follicles in which the eggs develop. Eggs are stored inside these follicles until they are ready to be released on their journey through the fallopian tubes (see photo in color insert), where they may ultimately be fertilized by sperm. Also inside the cortex are specialized cells that produce and secrete the steroid hormones estrogen and progesterone, as well as less potent male hormones (androgens). Ovarian sex hormones

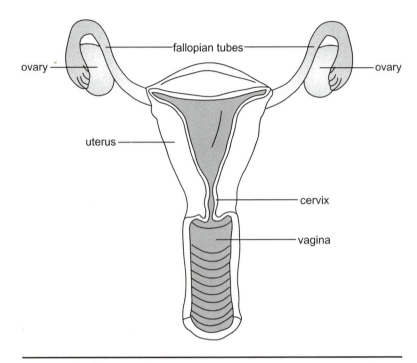

Figure 2.9. The ovaries.

are produced and released in response to follicle-stimulating hormone (FSH) and luteinizing hormone (LH) from the anterior pituitary. Once released, estrogens and progestins influence the development of the female reproductive organs and sexual characteristics. On the side of each ovary is a small notch, called the hilum, through which blood vessels and nerves enter and exit.

The two testes (see Figure 2.10) are located outside the man's body, suspended in a pouch called the scrotum. Their unusual location serves an important purpose: Inside the scrotum, they are kept a few degrees cooler than the internal body temperature to protect the sperm. The testes are made up primarily of **seminiferous tubules,** in which sperm are produced. In between the seminiferous tubules are about 350 million **Leydig (interstitial) cells,** which produce and release the hormone testosterone. Testosterone plays an important role in the development of male sexual characteristics and sexual organs (such as the prostate gland) and is essential to sperm production.

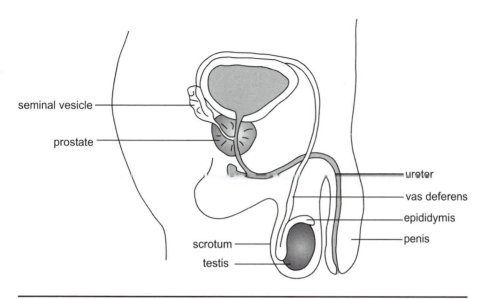

seminal vesicle

prostate

ureter

vas deferens

epididymis

penis

scrotum

testis

Figure 2.10. The testes.

NONENDOCRINE ORGANS THAT SECRETE HORMONES

Endocrine glands are not the only tissues capable of producing and releasing hormones. Scattered throughout the body are hormone-producing cells that are not technically part of the endocrine system. Cells within the heart, stomach, small intestines, and brain are all capable of synthesizing and secreting hormones.

Heart: Muscle cells (myocytes) in the heart contract and relax to pump blood through the body. These cells also produce two hormones: **atrial natriuretic peptide (ANP)** and **brain natriuretic peptide (BNP),** which act on the heart, adrenal glands, kidneys, lungs, and central nervous system. ANP and BNP are secreted in response to measured increases in heart rate and blood pressure. They function by increasing sodium and urine excretion by the kidneys, decreasing the heart rate, and relaxing the arteries.

Gastrointestinal System: The stomach and intestines not only digest and absorb nutrients from food, they also secrete a number of digestive hormones. Major gastrointestinal hormones include gastrin (regulates stomach acid secretion), **cholecystokinin** (triggers the release of digestive enzymes in the small intestine and stimulates gallbladder contraction), and **secretin** (stimulates pancreatic secretion during digestion).

Brain and Central Nervous System: Specialized nerve cells (neurosecretory cells) in the brain secrete neurohormones, which travel through the blood and trigger the release of endocrine hormones (see Chapter 3).

Vitamin D

Vitamin D, or **calciferol,** is found in food and is also made in the body. Exposure to ultraviolet light from the sun triggers synthesis of the inactive vitamin D precursor, vitamin D_3 (cholecalciferol), in the skin. The liver and kidneys then help metabolize vitamin D to its active form.

The primary purpose of vitamin D is to regulate calcium and phosphorous levels in the blood. It facilitates absorption of calcium in the intestines, and aids in bone mineralization to keep the bones strong.

Skin, Liver, and Kidneys: Vitamin D, a steroid hormone important in regulating calcium and phosphorous levels, is generated via a precursor molecule (**cholecalciferol**) from sunlight absorbed through the skin. Vitamin D is then metabolized into its active form in the liver and kidneys (see "Vitamin D").

Placenta: During pregnancy, the placenta nourishes the growing fetus. It also serves as a temporary endocrine gland, secreting hormones that ensure the viability of the pregnancy and step in for the developing fetal endocrine system until it is capable of producing its own hormones (see Chapter 7).

3

The Hypothalamus and Pituitary Gland

The hypothalamus (see photo in color insert) and pituitary gland together serve as the command center of the endocrine system, and the core of the relationship between the endocrine and nervous systems. Together, they regulate virtually every physiological activity in the body. As mentioned in Chapter 1, the nervous and endocrine systems also regulate each other: neurohormones from the hypothalamus direct the release of endocrine hormones, and hormones from the endocrine system regulate nervous system activity.

THE HYPOTHALAMUS

The hypothalamus (see Figure 3.1) is made up of clusters of **neurosecretory cells,** which both transmit electrical messages (impulses) and secrete hormones. Electrical impulses are transmitted from one nerve cell to another via chemical messengers called neurotransmitters. The impulses travel across junctions called synapses (see Figure 3.2) and bind to receptors on the receiving neuron.

Neurotransmitters are chemical compounds that are made up of simple or more complex amino acid sequences or peptides. Examples of neurotransmitters include epinephrine, norepinephrine, serotonin, **acetylcholine,** dopamine, and **histamine** (see Table 3.1).

The hypothalamus also secretes a number of hormones that are referred to as **neurohormones** (see Table 3.2), which either travel through the body via the general circulation or go directly to the anterior pituitary gland through

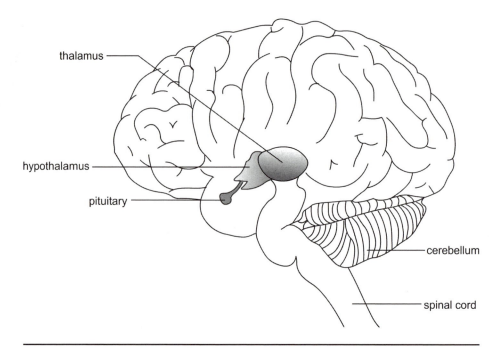

thalamus

hypothalamus

pituitary

cerebellum

spinal cord

Figure 3.1. The hypothalamus.

a portal network of blood vessels (called the hypothalamic-hypophyseal portal system) and signal it to release or stop releasing its hormones. Hypothalamic hormones are called releasing or inhibiting hormones, depending upon how they influence the pituitary gland. As their names suggest, releasing hormones trigger hormone secretion, and inhibiting hormones halt hormone secretion.

The following are hypothalamic neurohormones:

Thyrotropin-releasing hormone (TRH): This hormone is made of a simple three-amino-acid structure, but its function is relatively complex. TRH signals the pituitary gland to release thyroid-stimulating hormone (TSH) and prolactin (PRL). It also acts as a neurotransmitter in the brain and spinal cord. Some studies have indicated that TRH has therapeutic effects on the brain and spinal cord, mitigating the damage caused by spinal cord injury and lessening the severity of amyotrophic lateral sclerosis (Lou Gehrig's disease). TRH release is stimulated and inhibited by higher regions of the brain, and it is inhibited by high blood levels of thyroid hormone.

Gonadotropin-releasing hormone (GnRH) (also known as luteinizing hormone-releasing hormone [LHRH]): **Gonadotropin-releasing hormone (GnRH)** is a peptide chain of ten amino acids that stimulates release of the gonadotropins—luteinizing hormone (LH) and follicle-stimulating hormone (FSH)—from the

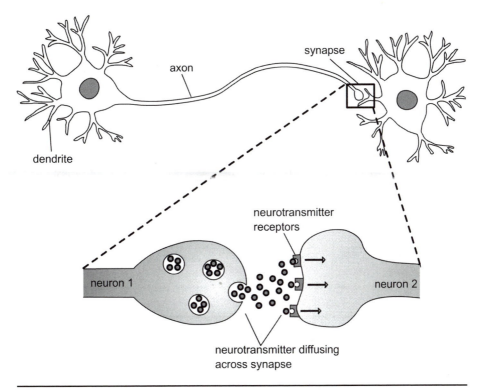

Figure 3.2. Neurotransmitter traveling across a synapse between neurons.

TABLE 3.1. Neurotransmitters

Neurotransmitter	Actions
Acetylcholine	Contracts or dilates muscles, increases stomach and urinary contractions.
Dopamine	Influences brain control of movement, emotion, and feelings of pain and pleasure.
Epinephrine (adrenaline)	Both a hormone and a neurotransmitter. Increases heart rate and blood pressure in response to stress.
Norepinephrine (noradrenaline)	Both a hormone and a neurotransmitter. Constricts blood vessels and increases heart rate in response to stress.
Serotonin	Stimulates smooth muscles, aids in transmission of electrical impulses across nerve cells, initiates sleep, elevates mood, and affects appetite.
Histamine	Involved in the immune response and allergic reactions.

TABLE 3.2. Hypothalamic Neurohormones

Neurohormone	Pituitary Hormone(s) It Affects
Thyrotropin-releasing hormone (TRH)	Thyroid-stimulating hormone (TSH) and pro-lactin (PRL)
Gonadotropin-releasing hormone (GnRH)	Luteinizing hormone (LH) and follicle-stimulating hormone (FSH)
Corticotropin-releasing hormone (CRH)	Adrenocorticotropic hormone (ACTH)
Growth hormone–releasing hormone (GHRH)	Stimulates growth hormone
Growth hormone–inhibiting hormone (GHIH) or somatostatin	Inhibits growth hormone, insulin and glucagon in the endocrine pancreas, and many functions of the gastrointestinal tract
Prolactin-stimulating and -releasing hormones	Releases and inhibits prolactin

anterior pituitary. LH and FSH act upon the gonads (the ovaries and testes). In females GnRH initiates the menstrual cycle, and in both sexes it helps determine the onset of puberty by triggering LH and FSH. The neurons that secrete GnRH are connected to a part of the brain known as the limbic system, which is involved in the controls of emotion and sexuality. GnRH release is stimulated by both negative and positive feedback by circulating levels of sex steroids. It is also stimulated by the neurotransmitter epinephrine and is inhibited by dopamine, endorphins, and melatonin.

Corticotropin-releasing hormone (CRH): **Corticotropin-releasing hormone (CRH)** is a large, forty-one-amino-acid peptide hormone that directs the body's stress response by stimulating the production and secretion of **adrenocorticotropic hormone** (ACTH) from the pituitary gland. In response to ACTH, the adrenal gland releases the stress hormone cortisol, which then suppresses CRH release (and subsequently its own release) via negative feedback. Antidiuretic hormone (ADH, or vasopressin) from the pituitary gland also plays a side role in stimulating CRH release.

Growth hormone–releasing hormone (GHRH): **Growth hormone-releasing hormone (GHRH)** is a large peptide hormone that stimulates the secretion of growth hormone from the anterior pituitary gland (see Figure 3.3). Release of GHRH is triggered by stress (such as exercise) and is inhibited by somatostatin, which is also released by the hypothalamus. Negative feedback is largely controlled by compounds known as **somatomedins,** growth-promoting hormones made when tissues are exposed to growth hormone.

Growth hormone–inhibiting hormone (GHIH), or somatostatin: This polypeptide hormone has the opposite effect of GHRH: It inhibits the release of growth hormone from the anterior pituitary gland (see Figure 3.4). GHIH also inhibits the release of thyroid-stimulating hormone (TSH). In the endocrine pancreas, GHIH serves a paracrine function by blocking the secretion of insulin and

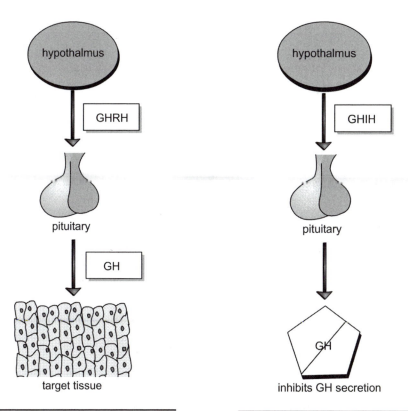

Figure 3.3. Hypothalamus releases GHRH, which travels to anterior pituitary to stimulate release of growth hormone.

Figure 3.4. Hypothalamus releases GHIH, which travels to anterior pituitary to inhibit release of growth hormone.

glucagon. It also acts upon the gastrointestinal tract, blocking the secretion of stomach acids and pancreatic enzymes, and influencing intestinal absorption.

Prolactin-inhibiting and-releasing hormones: The hypothalamus usually stimulates the release of pituitary hormones. But in the case of prolactin, it serves a primarily inhibitory role. The main prolactin inhibitor is the neurotransmitter dopamine. The main stimulator is thyrotropin-releasing hormone (TRH). A peptide produced in the hypothalamus, called **vasoactive intestinal peptide (VIP),** also stimulates PRL release.

THE PITUITARY GLAND

The pituitary gland (see Figure 3.5) is divided into two separate units: the anterior pituitary and posterior pituitary, each of which functions independently and secretes its own set of hormones.

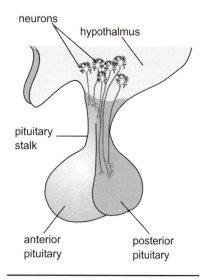

Figure 3.5. The pituitary gland.

The Anterior Pituitary

The anterior pituitary is made up of five different types of cells, each of which secretes one or more different hormones:

Thyrotroph—Thyroid-stimulating hormone (TSH)

Gonadotroph—Luteinizing hormone (LH) and follicle-stimulating hormone (FSH)

Corticotroph—Corticotropin (ACTH)

Somatotroph—Growth hormone

Lactotroph—Prolactin

Hormones are synthesized in the cytoplasm of the cell as larger, inactive molecules called prohormones. Neurohormones from the hypothalamus travel to the anterior pituitary via a closed system of veins (the hypothalamic-hypophyseal portal system) and signal the anterior pituitary to either release or stop releasing its hormones. When the signal is to release hormones, the hormone is activated from the prohormone as it is sent out from the cell into circulation. If the hypothalamus were destroyed, the anterior pituitary would be unable to secrete any of its hormones, with the exception of prolactin, which the hypothalamus primarily inhibits.

The anterior pituitary (see Table 3.3) affects growth and metabolism in most other endocrine glands, as well as in other areas of the body. It also stimulates other endocrine glands to produce and secrete their hormones:

THYROID-STIMULATING HORMONE (TSH)

Thyroid-stimulating hormone (TSH), also called thyrotropin, is a large glycoprotein that affects cell growth and metabolism in the thyroid gland, and signals the gland to produce and release its hormones thyroxine (T_4) and triiodothyronine (T_3). Thyrotroph cells in the anterior pituitary release TSH after being stimulated by TRH from the hypothalamus. TSH release is inhibited by negative feedback involving thyroid hormones. When blood levels of thyroid hormones are high, somatostatin inhibits the production of TRH from the hypothalamus, which then inhibits TSH release. Glucocorticoids and estrogens also serve an inhibitory function by making the pituitary less responsive to TRH.

GONADOTROPINS

Luteinizing hormone (LH) and follicle-stimulating hormone (FSH) are **gonadotropins** secreted by cells called gonadotrophs in the anterior pituitary. These hormones promote egg and sperm development and control hormone

TABLE 3.3 Anterior Pituitary Hormones

Pituitary Hormone	Acts Upon	Effects
Thyroid-stimulating hormone (thyrotropin)	Thyroid	Triggers release of thyroxine (T_4) and triiodothyronine (T_3)
Luteinizing hormone (LH) and follicle-stimulating hormone (FSH) (gonadotropins)	Ovaries and testes (gonads)	Triggers release of steroid hormones: estrogens, progestogens, androgens
Adrenocorticotropic hormone (ACTH) (corticotropin)	Adrenal cortex	Triggers release of glucocorticoids (cortisol), mineralocorticoids (aldosterone), and gonadocorticoids (androgens and estrogen)
Growth hormone	Liver, adipose tissue	Affects growth by influencing protein, fat, and carbohydrate metabolism
Prolactin	Mammary gland	Promotes development of breast tissue; initiates and sustains milk production during lactation

release by the ovaries and testes. The secretion of both hormones remains very low from shortly after birth until puberty, when levels rise dramatically (see Chapter 7). In females, the rate of secretion varies at different times during the menstrual cycle. Secretion of both LH and FSH is controlled by gonadotropin-releasing hormone (GnRH) from the hypothalamus.

Luteinizing hormone: This hormone was given its name because it stimulates the conversion of the ovarian follicle to the **corpus luteum** following egg release. In the middle of a woman's menstrual cycle, estrogen levels rise, causing the release of GnRH from the hypothalamus. GnRH causes LH levels to surge, triggering ovulation. That surge allows the egg to rupture from its follicle and travel down the fallopian tube toward the uterus. Once the egg has been released, LH stimulates the conversion of the ovarian follicle to the corpus luteum, which produces progesterone, a hormone necessary to maintain pregnancy. LH also stimulates estrogen and progesterone production. In men, the hormone stimulates the growth of—and testosterone production in—the Leydig cells of the testes.

Follicle-stimulating hormone: In women, FSH stimulates the maturation of the ovarian follicles, in which the eggs develop. With the help of LH, FSH also increases estrogen secretion by the ovaries. In men, FSH acts upon the **Sertoli cells** (cells that line the seminiferous tubules, which nourish the germ cells from which sperm develop) of the testes, facilitating sperm maturation and development. Like LH, the release of FSH is stimulated by GnRH from the hypothalamus. FSH is inhibited by the hormone **inhibin,** which is secreted by the ovaries and testes.

ADRENOCORTICOTROPIC HORMONE (ACTH)

Adrenocorticotropic hormone (ACTH), also called corticotropin, is a small peptide hormone that stimulates cell development and hormone synthesis in the adrenal cortex (glucocorticoids, mineralocorticoids, and gonadocorticoids). In the fetus, ACTH also stimulates secretion of an estrogen precursor called dehydroepiandrosterone sulfate (DHEA-S), which prepares the mother for labor. Stress stimulates the release of CRH from the hypothalamus. CRH then activates the secretion of ACTH. ADH (vasopressin) also plays a role in ACTH release. High circulating levels of cortisol in the blood inhibit ACTH in two ways: by directly suppressing ACTH synthesis and secretion in the pituitary, and by acting on the hypothalamus to decrease CRH release. ACTH may also inhibit its own secretion.

GROWTH HORMONE

Growth Hormone (GH), also called somatotropin, is a large polypeptide hormone produced by somatotroph cells in the anterior pituitary that plays a significant role in growth and metabolism. It primarily affects bone, muscle, and tissue growth. Without sufficient growth hormone, an individual would suffer from short stature (**dwarfism**; see Chapter 10). Too much growth hormone would result in **gigantism.** For normal growth to occur, the body requires energy, which growth hormone provides through protein synthesis and the breakdown of fats. Growth hormone has two types of effects: direct and indirect.

> *Direct effects*: Growth hormone acts directly upon protein metabolism, fat metabolism, and carbohydrate metabolism to help the body more efficiently use and conserve energy. It moves amino acids from the blood into cells and stimulates protein synthesis within the cells; it moves fats out of storage (in adipose tissue) for use in energy production to conserve proteins; and it decreases carbohydrate use and impairs glucose uptake into cells, thus sparing glucose for the brain.

> *Indirect effects*: Growth hormone stimulates bone, muscle, and cartilage growth indirectly, by triggering the production of insulin-like growth factor 1 (IGF-1, or somatomedin). **Insulin-like growth factors** are synthesized in the liver and other tissues and act much like insulin, stimulating glucose uptake by cells. They also influence protein and DNA synthesis. IGF-1 stimulates proliferation of cartilage cells (called **chondrocytes**), causes muscle cell differentiation and proliferation, and initiates protein synthesis in muscle tissues.

Growth hormone is released in pulses, which peak and ebb throughout the day. Secretion is highest in the early morning and lowest at midday. Several hormonal, dietary, and stress-related factors help regulate the intensity of growth hormone secretion:

Hypothalamic hormones: The hypothalamus both stimulates (via growth hormone–releasing hormone [GHRH]) and inhibits (via growth hormone–inhibiting hormone [GHIH or somatostatin]) the release of growth hormone. When blood levels of growth hormone or IGF-1 rise, they stimulate the hypothalamus to release somatostatin in a negative feedback loop.

Ghrelin: **Ghrelin,** a peptide hormone secreted by the stomach, is a potent stimulator of growth hormone secretion.

Metabolic controls: Carbohydrates, protein, and fat can all influence growth hormone secretion. High blood sugar (**hyperglycemia**) inhibits its release, while low blood sugar (**hypoglycemia**) stimulates its release. A surge in protein (for example, after a steak dinner) triggers growth hormone release, but, strangely, low protein in the blood also raises growth hormone secretion. This probably occurs because of a subsequent drop in IGF-1 production, which prevents the negative feedback loop that would otherwise inhibit growth hormone secretion. When fat is mobilized for energy (for example, during periods between meals), growth hormone release is stimulated to conserve protein.

Stress: Exercise and other stressors can stimulate growth hormone release. This response is believed to be part of the body's fight-or-flight mechanism, which mobilizes fat and glucose stores for energy.

PROLACTIN (PRL)

Prolactin (PRL) is a polypeptide produced not only in the lactotroph cells of the anterior pituitary, but also in the brain, the immune cells, and the cells of a pregnant woman's uterus. In females, the hormone has two primary roles:

Mammary gland development: During puberty, estrogen and progesterone stimulate breast development. During pregnancy, prolactin stimulates further development of the mammary glands to prepare them for milk production following childbirth.

Initiation and maintenance of milk production: Prolactin levels rise ten to twenty times normal amounts during a woman's pregnancy, yet milk secretion does not begin until the baby is born because it is inactivated by high circulating levels of estrogens and progestins. After the birth, levels of estrogens and progestins drop, enabling prolactin to initiate lactation (lactogenesis), which it does by affecting the genes that encode milk proteins. Prolactin also maintains milk production (along with oxytocin) until the infant is weaned. Even though prolactin levels drop in the mother about eight to ten weeks after delivery, they rise each time the infant suckles. That spurt of prolactin is enough to sustain milk production for as long as the infant nurses.

Prolactin secretion is controlled by prolactin-inhibiting and -releasing hormones from the hypothalamus. Its primary inhibitor is the hypothalamic neurotransmitter dopamine. Normally, dopamine suppresses prolactin secretion, but during lactation the stimulation of the mother's nipples when the baby nurses blocks the secretion of dopamine. Prolactin also inhibits the secretion of gonadotropin-releasing hormone (GnRH) from the hypothalamus, which stops LH and FSH from acting upon the ovaries. Without LH and FSH, the new mother does not menstruate and is unlikely to get pregnant again (although it can happen, especially when the baby begins to wean). Prolactin is usually not a factor in nonpregnant women, but it can sometimes be released following nipple stimulation.

Regulation of Anterior Pituitary Hormones

Hormones from the anterior pituitary can be regulated in one of three ways:

1. Hormones such as LH and FSH are released in pulses that follow a regular cycle. The strength and frequency of these pulses is set, in part, by the hypothalamus. Pulsatile release may follow a daily rhythm (circadian), or it may occur more frequently (ultradian) or less frequently (infradian) than once a day.

2. Most hormones are regulated by feedback loops, in which circulating hormone levels act upon the hormones that triggered their release. Three types of feedback loops exist:

   Long-loop system: After being stimulated by a releasing hormone from the hypothalamus (CRH), the anterior pituitary signals its target organ (for example, the adrenal cortex) to produce its hormone (cortisol) (see Figure 3.6). When that hormone reaches a certain level in the system, it acts upon the hypothalamus via negative feedback, inhibiting its releasing hormone (CRH). When the hypothalamus stops or decreases production of the releasing hormone, the anterior pituitary subsequently stops releasing its hormone (ACTH).
   Short-loop system: Some hormones (for example, LH and FSH) can suppress their own release without entering the bloodstream.
   Ultrashort-loop system: A releasing hormone (for example, LHRH or GHRH) can act directly on the hypothalamus to regulate its own secretion.

3. Finally, hormone release can be influenced by external factors, such as stress (for example, the fight-or-flight release of corticotropin from the adrenal cortex), diet, and illness.

The Posterior Pituitary

The posterior pituitary is not a classical endocrine organ because it is composed primarily of extensions of axons and nerve endings from the hypothalamus. Its two hormones, antidiuretic hormone (ADH, or vasopressin) and oxytocin, are actually produced in the neurons of nuclei in the hypothalamus. They travel down nerve fibers to the posterior pituitary, which merely stores and releases them (see Figure 3.7).

Antidiuretic hormone (ADH, or vasopressin) is a small peptide hormone whose primary role is to conserve water in the body by signaling the kidneys to excrete less fluid. The hormone is synthesized as a preprohormone in the hypothalamic neurons. The preprohormone is converted into a pro-

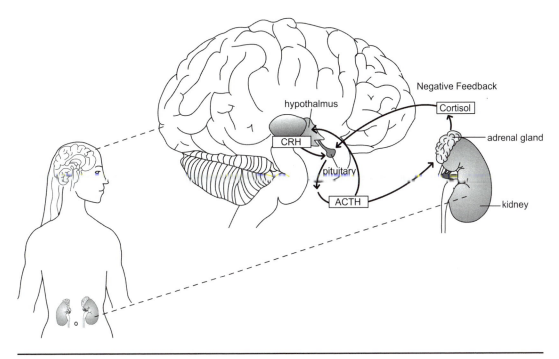

Figure 3.6. CRH-cortisol feedback loop.
CRH from the hypothalamus stimulates the anterior pituitary, which triggers the adrenal cortex to produce cortisol, which then acts upon the hypothalamus to inhibit CRH release.

hormone, which contains an attached protein called neurophysin that is removed as the hormone is secreted.

The human body contains about 60 percent water. As mentioned in Chapter 2, significant fluctuations in water balance (i.e., dehydration or overhydration) can be extremely dangerous to the system. When the concentration of liquid in the blood drops, special sensors (called osmoreceptors) in the hypothalamus alert the neurons that produce ADH. Osmoreceptors are extremely sensitive and can respond to tiny changes (as small as 1 percent) in water concentration.

Whereas a diuretic increases urine output, an antidiuretic conserves fluid

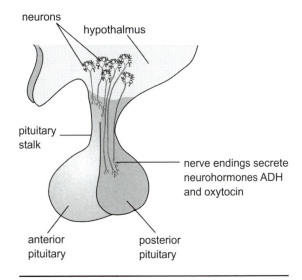

Figure 3.7. Hormones produced in neurons of the hypothalamus travel down nerve fibers to the posterior pituitary.

in the body by reducing urine output. When ADH is released, it binds to receptors in the distal or collecting tubules of the kidneys and increases their permeability, thus stimulating the reabsorption of liquid back into the blood (normally, these tubules are virtually impermeable to water). When more water is absorbed into the bloodstream, blood volume and pressure increase. Conversely, when the fluid level in the body gets too high, ADH release is suppressed, and the kidneys excrete more liquid into the urine. ADH also constricts blood vessels (vasoconstriction), the role for which it was given its alternate name, vasopressin.

Thirst, the body's physical indicator that fluid levels are low, is also regulated by osmoreceptors in the hypothalamus, although not the same ones that trigger ADH release. The body first sends in ADH to try to regulate water balance, then, if that measure fails to increase fluid volume, it invokes thirst.

Changes in blood pressure also stimulate ADH release. Two pressure sensors—one in the carotid artery in the neck, and the other in specialized cells in the atrium of the heart—discern changes in blood pressure and volume (see Figure 3.8). They send a message, via nerves, to the hypothalamus. When these sensors are stretched by expanding blood volume, they shut off

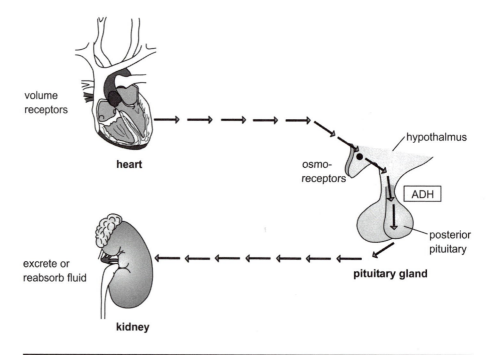

Figure 3.8. Blood pressure and ADH secretion.
Volume receptors in the heart sense changes in blood pressure and send a message to the hypothalamus, which either stimulates or inhibits ADH secretion to regulate fluid absorption by the kidneys.

ADH secretion so that more water is excreted by the kidneys. When they sense reduced blood volume (for example, when a person is hemorrhaging from a severe injury), they trigger ADH production.

In addition to regulating fluid volume, ADH has several peripheral effects:

- Inhibits renin secretion

- Stimulates ACTH secretion

- Is thought to influence behavior and facilitate memory and learning

- Plays a role in regulating the cardiovascular system

Oxytocin

Oxytocin is similar in structure to ADH, and it is also synthesized from preprohormones in the hypothalamus and transported to the posterior pituitary for secretion into the blood. In addition to being secreted by the pituitary, oxytocin is also released by tissues in the ovaries and testes. The primary function of oxytocin is to stimulate the mammary glands in a mother's breast during lactation—to let down the milk so that her baby can nurse. Oxytocin (which is derived from the Greek word meaning "swift birth") also stimulates uterine contractions during labor. And the release of oxytocin into a new mother's brain helps forge a bond between her and her new baby. The hormone normally circulates in low levels in both men and women, but it rises in women during ovulation, birth, and lactation, as well as in times of stress.

MILK PRODUCTION

The hormone prolactin (see "Prolactin [PRL]" in this chapter) stimulates milk production in the mother's breasts during pregnancy and lactation. Milk is secreted into small sacs called **alveoli** (not to be confused with the structures in the lungs of the same name). The alveoli are surrounded by smooth muscle cells called myoepithelial cells. Oxytocin, which is released when the child is born, targets those muscle cells and stimulates them to contract (see Figure 3.9). The contraction sends milk into the nipple.

Oxytocin release is stimulated externally by the touch of the nursing infant. The suckling activates special receptors in the nipple that send a signal to the hypothalamus, triggering hormone secretion. Oxytocin release may also occur via uterine or cervical stimulation during sex (which is why nursing mothers sometimes leak milk during intercourse), or it may be released without any physical stimulation but merely at the sight or sound of a crying infant. Pain, stress, or fear can all block oxytocin release.

UTERINE CONTRACTIONS DURING LABOR

When the vagina and cervix stretch during the onset of labor, oxytocin stimulates the smooth muscle cells surrounding the mother's uterus to con-

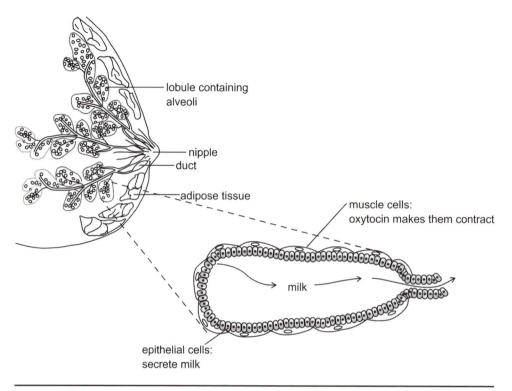

Figure 3.9. Oxytocin and lactation.
Oxytocin stimulates muscle cells surrounding the alveoli. The contraction sends milk into the nipple.

tract, which helps push the baby out. Doctors sometimes give women a synthetic form of oxytocin, called **pitocin,** to stimulate uterine contractions when labor is not progressing appropriately (this is called induced labor). Pitocin stimulates forceful contractions almost immediately after it is administered.

ASSISTS IN REPRODUCTION

There is also evidence that oxytocin assists in reproduction by facilitating sperm transport through the male—and female—reproductive tracts.

MATERNAL BONDING

Following childbirth, a rush of oxytocin to the new mother's brain provides a calming effect and helps her bond with her new baby.

PINEAL GLAND

The pineal gland in the center of the brain (see Figure 3.10) is filled with specialized cells called **pinealocytes,** which produce the hormone melatonin. Melatonin synthesis begins with the amino acid tryptophan, which

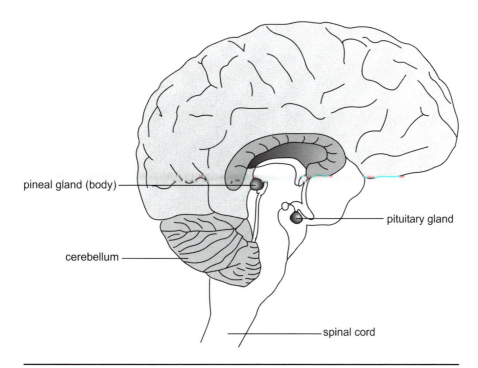

pineal gland (body)

pituitary gland

cerebellum

spinal cord

Figure 3.10. The pineal gland.

is then converted into serotonin and finally into melatonin. Melatonin release is regulated by the sympathetic nervous system and is stimulated primarily by darkness, but it can also be triggered by hypoglycemia (low blood sugar). Melatonin concentration is highest at night and falls to almost undetectable levels during the day.

Although its function is still not completely clear, melatonin is believed to act upon suprachiasmatic nuclei (tightly-packed groups of small cells) in the hypothalamus (which have receptors to it) to influence the body's daily biological rhythms. Synthetic versions of the hormone have been used to treat everything from jet lag to insomnia.

The Thyroid and Parathyroid Glands

The thyroid and parathyroid glands in the neck have several life-sustaining functions: The thyroid gland produces hormones that affect growth, development, metabolism, calcium homeostasis, and cell differentiation, and the parathyroid glands regulate calcium and phosphorous levels.

THE THYROID GLAND

The thyroid gland (see Figure 4.1), the largest endocrine organ, is crucial to nearly all of the body's physiological processes. It produces thyroid hormones, which are needed for growth, development, and a variety of metabolic activities. The thyroid is composed of two types of cells: follicular and parafollicular (see photo in color insert).

Follicles are sacs filled with the prohormone thyroglobulin. Thyroglobulin breaks apart to produce the two thyroid hormones, thyroxine (T_4) and triiodothyronine (T_3). Lining the sacs are follicular cells, which synthesize and then either secrete or store these hormones. Parafollicular (or C) cells fill the spaces in between follicles. They secrete the hormone calcitonin. Inside the follicles is a substance called colloid, which is primarily made up of the glycoprotein thyroglobulin.

Thyroid Hormone Synthesis

What makes the thyroid gland unusual among endocrine organs is that it requires iodine to produce its hormones. Iodine enters the body through food (i.e., iodized salt and bread) and water in the form of iodide or iodate ion.

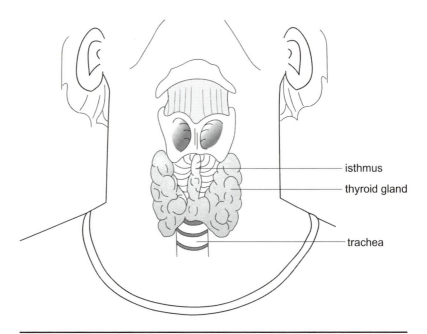

Figure 4.1. The thyroid gland.

The thyroid gland captures this iodide, and the enzyme thyroid peroxidase activates it (see Figure 4.2). The follicular cells produce thyroglobulin, which is deposited in the colloid. Tyrosine is bound to the thyroglobulin molecule. Iodine diffuses into the colloid and is added to the thyroglobulin. Enzymes break down the thyroglobulin, releasing thyroid hormones into the bloodstream.

Without sufficient iodine, the thyroid reduces its hormone output. Decreased levels of thyroid hormones in the blood stimulate the anterior pituitary to secrete more thyroid-stimulating hormone (TSH) to make up for the deficit. The thyroid gland swells in size as it tries to increase its output, a condition called goiter (see Chapter 11).

Unlike most other endocrine organs, which produce and immediately secrete their hormones, the thyroid can store its hormones for several weeks. It releases its hormones when acted upon by thyroid-stimulating hormone (TSH) from the pituitary gland. TSH stimulation leads to the activation of thyroxine (T_4). It splits from the thyroglobulin molecule as it leaves the cell and enters the bloodstream. T_4 is the more plentiful of the two hormones, making up about 90 percent of the total thyroid hormones. Triiodothyronine (T_3) is secreted in much smaller concentrations than thyroxine, but it is the more active of the two hormones. T_3 is produced (usually in the pe-

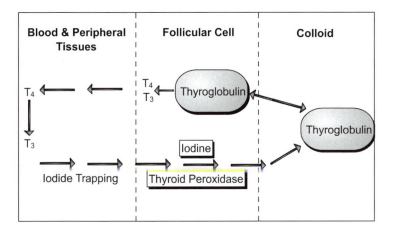

Figure 4.2. Thyroid hormone synthesis.
(1) The thyroid traps iodide, and thyroid peroxidase activates it; (2) in the colloid, iodine is added to the thyroglobulin molecule; (3) enzymes break down the thyroglobulin, releasing T_4 and T_3 into the bloodstream.

ripheral tissues, especially the liver and kidney) when thyroxine loses one of its iodine molecules.

Because thyroid hormones are not water soluble, they generally travel through the bloodstream attached to carrier proteins (most often thyroxine-binding globulin, but also to a lesser extent to thyroxine-binding prealbumin and albumin). Virtually every cell in the body contains thyroid hormone receptors. Once the thyroid hormone reaches its target cells, it travels through the membrane via diffusion or with the help of carriers, and it binds to receptors in the nucleus. Thyroid hormones affect gene transcription, which either stimulates or inhibits protein synthesis.

Thyroid Hormone Effects

As mentioned, thyroid hormones act upon virtually every cell in the body. Because their primary role is to regulate metabolism, an excess or lack of thyroid hormones can impair virtually every metabolic function, including growth, reproduction, and energy balance. Individuals who suffer from **hyperthyroidism,** an overproduction of thyroid hormones, experience insomnia, fatigue, irritability and nervousness, irregular heartbeat, increased sensitivity to heat, frequent bowel movements, and an increased metabolism. **Hypothyroidism,** a deficiency in thyroid hormones, is characterized by weakness, fatigue, depression, weight gain, constipation, increased sensitivity to cold, and decreased libido (more on these conditions in Chapter 11).

Other effects of thyroid hormones include the following:

Metabolism. Thyroid hormones regulate the body's energy consumption—the rate at which it uses fuel it receives from food. These hormones increase cellular oxygen use and heat production and stimulate protein synthesis, fat mobilization, and carbohydrate metabolism.

Growth. Thyroid hormones are necessary for skeletal growth. Children who do not produce enough thyroid hormones typically suffer from stunted growth, because these hormones are necessary for bone formation and maturation (see Chapter 11).

Brain and Central Nervous System. Thyroid hormones are also essential for brain and central nervous system development. In the fetus, thyroid hormones facilitate neuron proliferation and differentiation, and aid in synapse formation. If the fetal thyroid does not work properly, the child may be born with severe mental retardation. The thyroid continues to play an important role in brain and central nervous system function into adulthood. Individuals with hypothyroidism may feel sluggish or lethargic. People with hyperthyroidism are often irritable, anxious, and nervous, and may suffer from insomnia.

Reproductive System. Adequate levels of thyroid hormones are essential for normal reproductive function. Hypothyroidism can slow or even stop menstruation, often leading to infertility. Women who develop hypothyroidism during pregnancy are at increased risk for preeclampsia (high blood pressure), placental abruption (premature separation of the placenta from the uterus), premature delivery, or stillbirth.

Cardiovascular System. Thyroid hormones influence heart rate and blood flow. When they are under- or overproduced, the result can be dangerous fluctuations in heart rate and blood pressure. Patients with hyperthyroidism may experience increased heart rate, an enlarged heart, and increased systolic blood pressure. Those who suffer from hypothyroidism often have unhealthily high levels of low-density lipoproteins (LDL, or bad cholesterol), a risk factor for cardiovascular disease.

Gastrointestinal System. Because thyroid hormones stimulate gut motility (gut muscle contractions that push digested food through), hyperthyroidism can cause diarrhea, and can lead to constipation.

Regulation of Thyroid Hormone Secretion

The hypothalamic-pituitary axis is the primary regulator of thyroid hormone production and secretion (see Figure 4.3). Thyrotropin-releasing hormone (TRH) from the hypothalamus travels to the anterior pituitary and stimulates it to release thyroid-stimulating hormone (TSH). TSH influences every step of the thyroid production process, from thyroid cell growth to iodide uptake and metabolism. Finally, TSH triggers hormone secretion.

TSH release is stimulated and inhibited by positive and negative feedback, based on circulating levels of T_4 and T_3. When these hormones are in short supply, the pituitary and hypothalamus act on the thyroid to increase its production. Conversely, when too much of these hormones circulate in

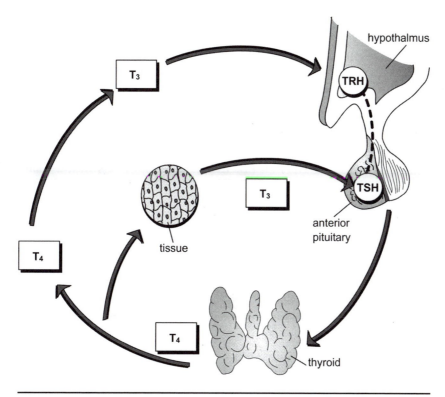

Figure 4.3. Thyroid hormone regulation.
(1) TRH from the hypothalamus triggers TSH secretion from the pituitary, (2) TSH stimulates thyroid hormone release, and (3) circulating levels of T_4 and T_3 influence the hypothalamus and pituitary to regulate thyroid production.

the blood, the pituitary and hypothalamus slow or stop thyroid production. Feedback ensures that hormone levels remain at an appropriate level.

Also controlling thyroid hormone secretion is the conversion of T_4 into T_3. As more T_4 is converted into T_3, rising T_3 levels make the pituitary less responsive to TRH. But as more T_4 is lost because of this conversion, the pituitary once again becomes more sensitive to TRH stimulation.

Thyroid hormone secretion can also be influenced by several other mechanisms:

Iodide levels: The thyroid may regulate its own secretion, without stimulation by TSH, based on iodide levels in the body. When the supply diminishes, for example, the thyroid must adapt by conserving and using iodide more efficiently.

Immune system: The immune system can stimulate or inhibit thyroid function. White blood cells called B lymphocytes produce TSH-receptor antibodies, which can act like TSH to trigger thyroid hormone release or can block their release.

External influences: Cold and other external influences may trigger TRH secretion. When the body is cold, for example, the thyroid secretes more of its hormones to raise body heat.

CALCIUM AND PHOSPHOROUS HOMEOSTASIS

The thyroid and parathyroid glands work together to carefully regulate calcium balance. Calcium is absorbed from ingested foods in the small intestine. Most of the calcium (99 percent) in the body is held in the bones. The tiny remainder circulates in the extracellular fluid. Calcium moves continually between the intestines, bones, and blood (see Figure 4.4). The body

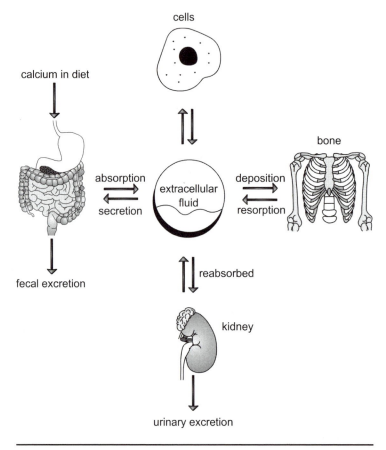

Figure 4.4. Movement of calcium between intestines, bones, and blood.

must maintain a balance so that calcium levels remain constant even as calcium is ingested from food and removed via the urine and feces.

When blood calcium levels are too high, calcium exits the circulation and enters the bone. Also, more calcium is excreted in the urine. Conversely, when blood calcium levels are too low, calcium moves from the bone into the blood and less is excreted in the urine.

Calcium balance is essential because the element is crucial to a number of body functions. It ensures that muscles (including the heart) contract properly. It maintains skeletal strength and structure. It is necessary for blood clotting. It is involved in hormone secretion and cellular response to hormones, and it acts as a second messenger for many hormones. Finally, it assists in the transmission of nerve impulses across synapses.

Because calcium is so essential to proper body function, its concentration in the blood and bones is carefully regulated by the hormones calcitonin (from the thyroid), parathyroid hormone (PTH—from the parathyroid glands), and the hormone-like vitamin D. The release of these substances occurs based on very subtle changes in blood calcium levels.

Calcitonin

Calcitonin, a peptide hormone, is synthesized and secreted by the parafollicular (C) cells in the thyroid gland. It is also synthesized in other tissues, including the lungs and intestines. How calcitonin is produced and synthesized remains, for the most part, a mystery.

Calcitonin's main job is to prevent calcium from leaving the bones and entering the blood, thus lowering blood calcium levels. Bones are continually renovating themselves in a process called "remodeling." Cells called **osteoclasts** remove old bone; **osteoblasts** build new bone (see Figure 4.5). Calcitonin slows the formation of osteoclasts and inhibits their ability to remove old bone, while it promotes the actions of osteoblasts. Calcitonin also releases phosphorous (another key component of bone) from the bones into the blood. In the kidneys, calcitonin stops the reabsorption of calcium and phosphorous back into the bloodstream, which means that more of these substances are excreted in the urine.

Calcitonin Regulation

The thyroid releases calcitonin whenever blood levels of calcium rise too high. Its secretion continues until blood calcium levels return to normal. Calcitonin release works in opposition to parathyroid hormone (PTH), which is released when blood calcium drops too low (see below). When calcium concentration rises, PTH release is inhibited and calcitonin release is stimulated.

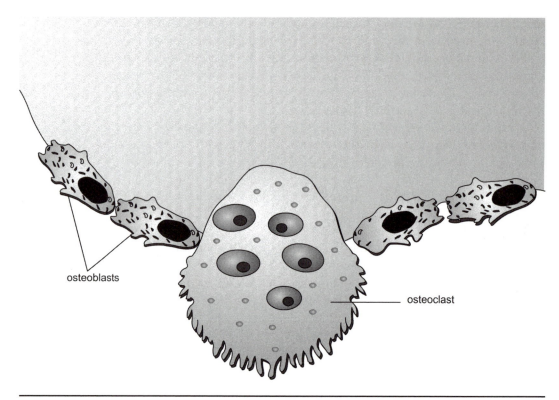

Figure 4.5. Osteoclasts and osteoblasts in bone.
Calcitonin promotes osteoblast activity, while PTH stimulates osteoclasts.

THE PARATHYROID GLANDS

The parathyroid glands (see Figure 4.6) produce only one major hormone: parathyroid hormone (PTH).

Parathyroid Hormone

Parathyroid hormone (PTH; also called parathormone), a polypeptide, opposes the actions of calcitonin by increasing blood calcium levels. Without this hormone, calcium concentrations in the blood would drop to life-threatening levels (**hypocalcemia**). Another effect of PTH is to reduce blood levels of phosphorous.

The parathyroids synthesize PTH from a larger, inactive prohormone in the parenchymal parathyroid cell. As PTH is released from the cell, it is split from the prohormone.

PTH increases blood calcium by acting upon the bones, the kidneys, and (indirectly) the small intestine:

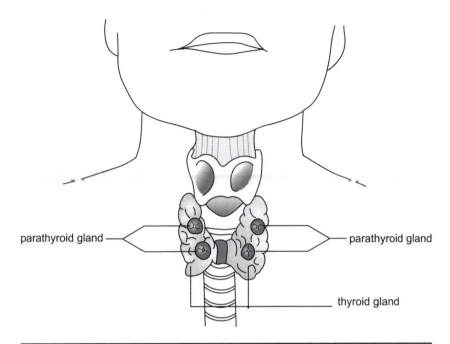

parathyroid gland

parathyroid gland

thyroid gland

Figure 4.6. The parathyroid glands.

Bone: PTH releases calcium from bone by stimulating the formation and activity of bone-dissolving cells called osteoclasts.

Kidneys: PTH increases calcium reabsorption in the kidney tubules, reducing the amount of calcium that is lost in the urine. Because the kidneys filter a large volume of calcium each day, even a slight adjustment in excreted calcium can have a big effect on body chemistry. PTH also decreases phosphorous reabsorption, so the kidneys excrete a greater amount.

Small Intestine: The effects of PTH on the small intestine are indirect. PTH increases production of vitamin D metabolites (the active form of vitamin D) in the kidneys (more on vitamin D later in this chapter). These metabolites increase the rate at which ingested calcium is absorbed in the small intestine, providing more calcium to circulate in the bloodstream.

PTH Regulation

Because blood calcium levels are so crucial to normal body function, cells of the parathyroid contain special receptors that can sense minute changes in calcium concentration. When calcium binds to these receptors, it results in reduced PTH secretion, which lowers calcium concentration in the blood (see Figure 4.7). Without the influence of bound calcium, the receptors con-

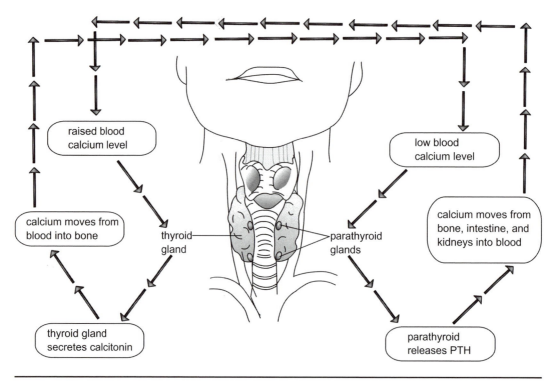

Figure 4.7. Calcitonin and PTH regulation.

tinue to stimulate PTH secretion. In the event that blood calcium levels remain depressed, PTH secretion can increase to fifty times its normal levels.

Parathyroid Hormone–Related Protein (PTHrP)

Parathyroid hormone–related protein is similar in structure and function
to parathyroid hormone (it too affects calcium and phosphorous balance),
but it is produced in many tissues throughout the body. PTHrP binds to PTH
receptors and has several of its own receptors as well. PTHrP can either act
upon cells in other parts of the body or influence the nucleus of the cell(s)
in which it was produced (called intracrine action). Like PTH, it releases
calcium from bone into the blood and increases reabsorption in the kidneys.

In addition to its role in calcium transfer, PTHrP has several other peripheral roles:

- Influences breast development and lactation

- Relaxes smooth muscles in the bladder, uterus, and heart

- Involved in tissue and organ development (including hair follicles and
  teeth)

Vitamin D

Vitamin D (also known as calciferol) is categorized as a vitamin, although it acts more like a hormone. Vitamin D enters the body through foods (including fish oils, fish liver, eggs, and vitamin-fortified products) ingested in the intestine, or via the metabolism of sunlight through the skin. It begins life as an inactive prohormone, most commonly cholecalciferol (vitamin D_3). Cholecalciferol is bound to a protein and transported to the liver and kidneys, where it is metabolized into its active form.

Vitamin D works in conjunction with PTH to increase blood levels of calcium and phosphorous by increasing calcium and phosphorous absorption in the intestines, making more of these substances available to bones; promoting bone formation and turnover; and stimulating calcium reabsorption in the kidney tubules.

Scientists believe that vitamin D also influences a number of other organs and tissues, including the liver (regeneration), immune system (cytokine production), muscles (contraction), and endocrine system (production and secretion of insulin, prolactin, and PTH).

Vitamin D Regulation

PTH and vitamin D production and secretion are tightly integrated to maintain calcium and phosphorous homeostasis. If blood levels of calcium and (to a lesser extent) phosphorous drop, PTH levels rise. PTH triggers the release of calcium and phosphorous from bone and the retention of calcium by the kidneys. PTH synthesis also facilitates the production of active vitamin D metabolites in the kidneys. Rising blood calcium levels inhibit PTH and therefore vitamin D production.

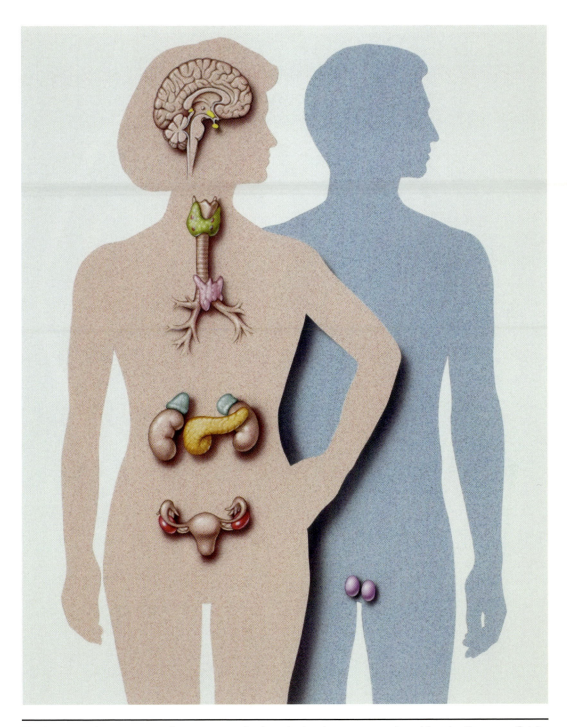

The human endocrine system showing major glands and related organs on male and female silhouettes. © John Karapelou, CMI/Phototake.

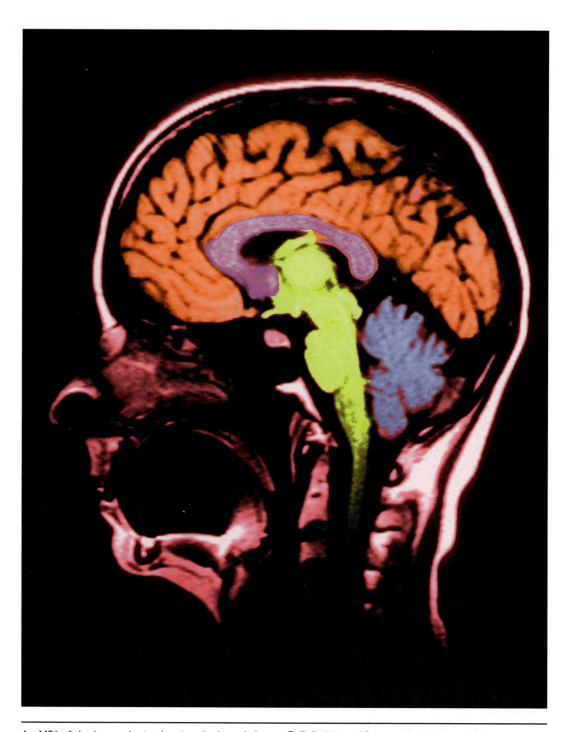

An MRI of the human brain showing the hypothalamus. © D. R. Unique/Custom Medical Stock Photo.

Follicular cells line the thyroid follicles, parafollicular cells fill the spaces between the follicles, and colloid fills the interior of the follicles. © Educational Pictures/Custom Medical Stock Photo.

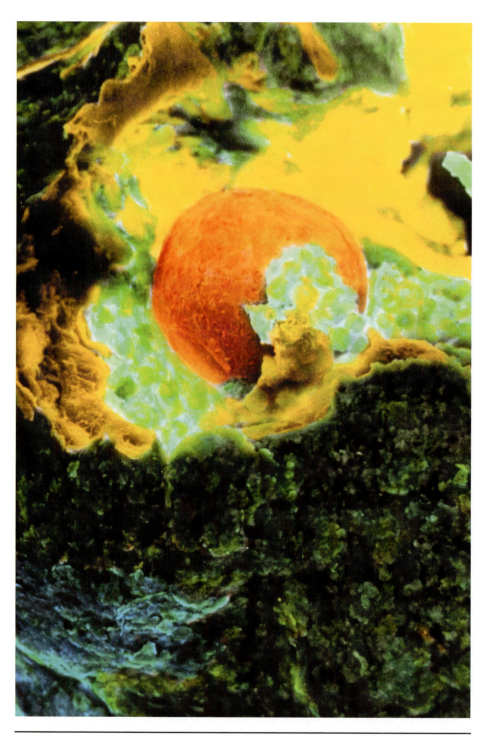

The ovulation process, showing a ruptured egg. © Science Photo/Custom Medical Stock Photo.

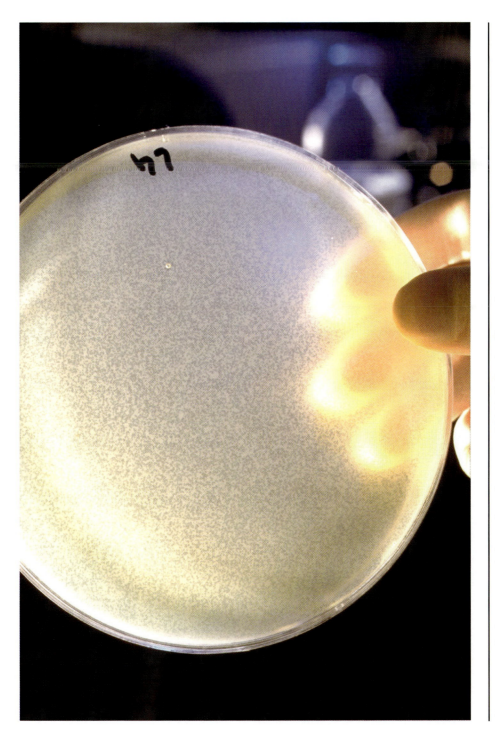

Obesity research: the process of cloning large fragments of DNA. © Eurelios/Phototake.

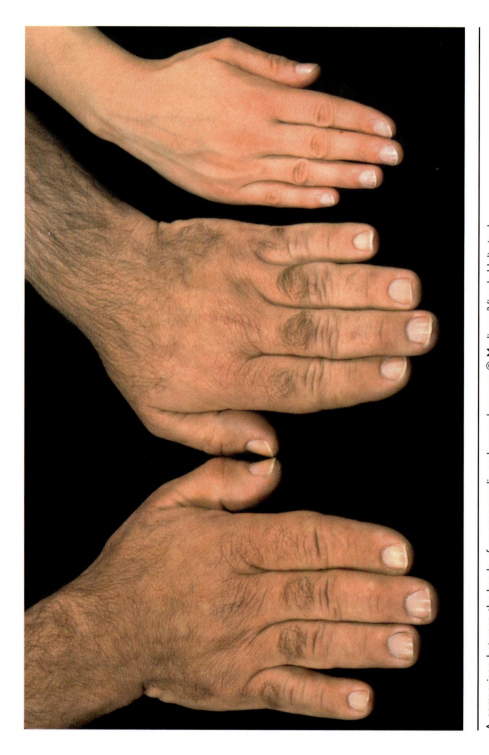

A comparison between the hands of an acromegalic and normal person. © Mediscan/Visuals Unlimited.

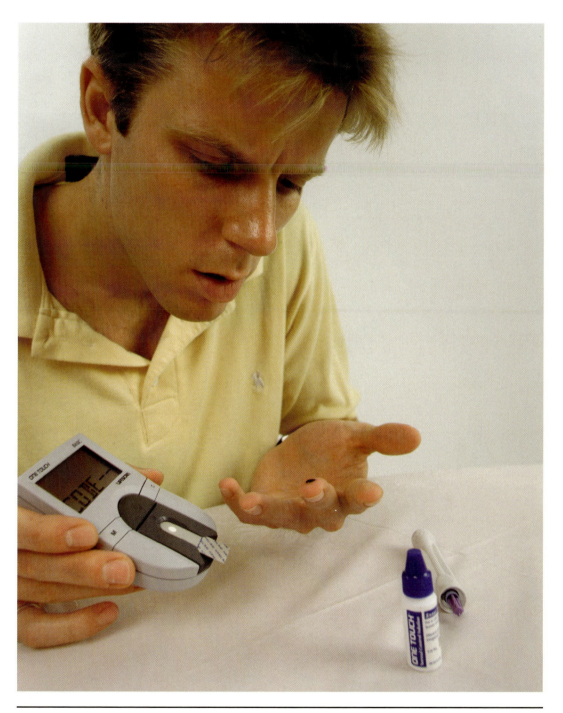

A young man testing his blood sugar using a glucometer. © Yoav Levy/Phototake.

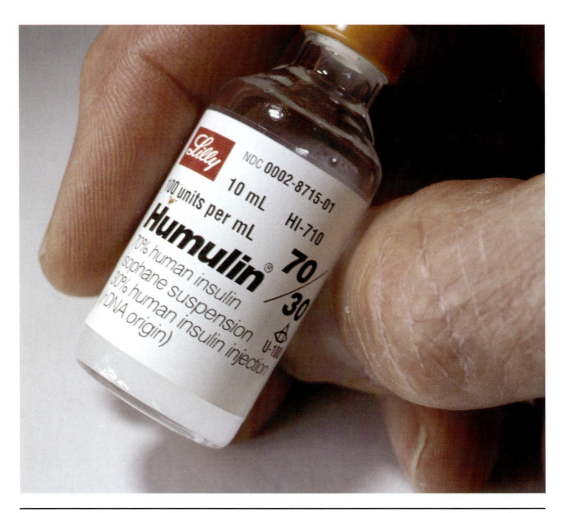

Human insulin for the control of diabetes, made by using recombinant DNA. © David Wrobel/Visuals Unlimited.

The Adrenal Glands and Endocrine Pancreas

When the body is confronted with stress—a serious car accident or a career-ending confrontation with one's boss, for example—the two adrenal glands kick into gear. The substances they produce (the catecholamines from the adrenal medulla and the steroid hormones from the adrenal cortex) orchestrate the stress response, making the body more alert, more energy efficient, and ready to face the daunting task at hand. Adrenal hormones are also involved in a number of other functions: regulating electrolyte balance, blood sugar levels, and metabolism, and influencing sexual characteristics.

ADRENAL MEDULLA

Hormones produced in the central, medullary region of the adrenal gland are referred to as the catecholamines. They are both hormones and neurotransmitters, because they are produced and secreted by sympathetic nerves (including neurons in the brain). The primary catecholamines are epinephrine (adrenaline), norepinephrine (noradrenaline), and dopamine.

Catecholamines are produced in a multistep process that begins with the amino acid tyrosine. First, an enzyme converts tyrosine into the chemical L-dopa. A second enzyme converts L-dopa to dopamine. Dopamine is then converted to norepinephrine by yet another enzyme. Finally, epinephrine is synthesized from norepinephrine before being released into the blood. Epinephrine makes up the bulk (80 percent) of catecholamines released from the adrenal medulla; the remaining 20 percent is norepinephrine.

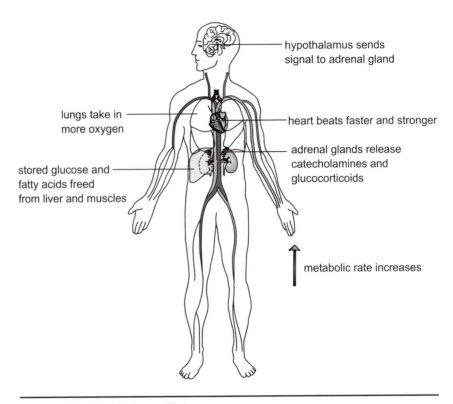

hypothalamus sends
signal to adrenal gland

lungs take in
more oxygen

heart beats faster and stronger

adrenal glands release
catecholamines and
glucocorticoids

stored glucose and
fatty acids freed
from liver and muscles

metabolic rate increases

Figure 5.1. The fight-or-flight response.

Catecholamine release is part of what Harvard physiologist Walter Cannon (1871–1945) termed the body's fight-or-flight response (see Figure 5.1). When confronted with stress, the body shifts into overdrive. It becomes more alert and refocuses energy as it prepares to either stay and fight the danger—or run away as quickly as possible.

As soon as the physical or emotional trauma occurs, the hypothalamus sounds the alarm. It sends out nerve impulses, which race to the adrenal medulla and signal it to release epinephrine and norepinephrine. These substances course through the blood attached to carrier proteins such as albumin, and they bind to adrenergic receptors on the surface of their target cells. Two types of adrenergic receptors exist: alpha-adrenergic and beta-adrenergic. The response elicited depends on the type of receptor stimulated. Alpha receptors are involved in smooth muscle contractions, pupil dilation, and blood vessel contraction. Beta receptors stimulate the heart and lungs, and relax the uterus.

Once the hormones are bound to their receptors, the body undergoes a rapid and dramatic transformation. The heart beats faster and stronger, the pupils dilate, the skin breaks out in a sweat, and the breathing becomes more intense as the body reaches a new level of alertness. Under the surface, a number of important physiological changes occur, directed by the adrenal catecholamines:

- The heart beats faster and more forcefully (primarily as a result of epinephrine), rushing additional blood throughout the body (especially to the brain and muscles). At the same time, the arteries constrict (primarily as a result of norepinephrine), increasing blood pressure.

- Stored glucose and fatty acids are freed to be used for energy. Catecholamines release glucose from the liver and muscles by stimulating the breakdown (**glycogenolysis**) of **glycogen**—the stored form of glucose. They also release fatty acids from **adipose tissue** by breaking down fatty compounds called triglycerides. The catecholamines oppose the action of insulin by preventing glucose movement into muscle and adipose tissue. Glucose is therefore preserved for use by the brain, which needs it most.

- The metabolic rate increases: Oxygen consumption and body heat rise.

- In the lungs, small tubules called bronchioles dilate, increasing the flow of air.

- Smooth muscles in the gastrointestinal tract and sphincters contract, while muscles in the uterus and trachea relax.

- Motor activity, gastrointestinal secretion, and other nonessential activities slow to conserve energy for other, more crucial functions.

Dopamine, a neurotransmitter, is similar to epinephrine. It influences the brain processes controlling emotion, movement, and the sensations of pleasure and pain. When dopamine is not produced in large enough quantities (for example, in patients with Parkinson's disease), the body grows rigid and movement becomes difficult.

ADRENAL CORTEX

The large outer region of the adrenal gland, the cortex (see Figure 5.2), is made up of three layers or zones, each of which produces its own group of steroid hormones:

Zona glomerulosa, the outermost layer, is where the mineralocorticoids (aldosterone) are produced.

Zona fasciculata, the middle layer, is where the glucocorticoids (cortisol) are produced.

Zona reticularis, the innermost layer, is where the gonadocorticoids (sex hormones androgens and estrogens) are produced.

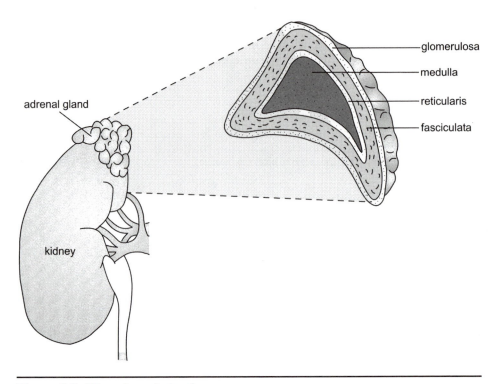

Figure 5.2. The adrenal gland.
The medulla is the central region of the adrenal gland. It is surrounded by the cortex, which is divided into the three zones: the reticularis, fasciculata, and glomerulosa.

The adrenal cortex is separated into two functional regions, each regulated by a separate entity. The fasciculata and reticularis layers depend upon adrenocorticotropic hormone (ACTH) stimulation from the anterior pituitary. Without ACTH, these two regions would atrophy. The zona glomerulosa is under the control of the renin-angiotensin system, which regulates blood pressure (more on this later in the chapter).

Mineralocorticoids, glucocorticoids, and gonadocorticods are all steroid hormones derived from cholesterol. Most of the cholesterol that enters the adrenal gland comes from **low-density lipoproteins (LDLs)** circulating in the blood. Upon stimulation of the adrenal cortex, cholesterol that splits off from the LDL is converted into pregnenolone, the precursor molecule for steroid hormones. The adrenal gland can also synthesize a small amount of its own cholesterol.

Mineralocorticoids

The principal mineralocorticoid, aldosterone, is produced and secreted by the zona glomerulosa. Aldosterone acts upon the kidneys to regulate sodium, potassium, and water reabsorption. It stimulates the distal tubules to reabsorb more sodium and excrete more potassium (as well as hydrogen) in the urine. As sodium is reabsorbed, more water is also reabsorbed, increasing blood fluid volume. Aldosterone also acts upon the salivary glands, sweat glands, and colon to reduce the amount of sodium lost in saliva, sweat, and feces.

Sodium and potassium balance is crucial, because these fluids regulate fluid movement between the cells and the extracellular fluid. Without aldosterone, a deadly fluid imbalance could result. When too much sodium is present, water from inside the cells crosses over into the extracellular region to restore balance, causing the cells to shrink—a situation that could lead to shock and death. Too little sodium can send water into the cells, causing them to swell and potentially leading to nausea, vomiting, diarrhea, convulsions, or coma.

Aldosterone release is primarily under the control of the renin-angiotensin system (see Figure 5.3), which helps regulate blood volume and blood pressure. Inside the blood vessels of the kidneys are tiny receptors that can detect changes in blood pressure and extracellular fluid volume. When these sensors notice a drop in pressure and volume, they release the enzyme renin into the blood. Renin travels to the liver, where it converts the protein **angiotensinogen** into another protein, called **angiotensin I.** Once angiotensin I reaches the lungs, it is converted by angiotensin-converting enzyme (ACE) into the much more potent hormone, **angiotensin II,** which constricts the blood vessels and stimulates the zona glomerulosa to synthesize aldosterone, raising blood pressure.

Blood potassium and sodium levels can also influence aldosterone secretion. For example, after a meal of avocado salad and banana pudding (both of which are high in potassium), aldosterone secretion increases. Aldosterone signals the kidneys to excrete more potassium until levels fall to within the normal range. Conversely, a potassium deficiency suppresses aldosterone secretion. Sodium has the opposite effect on aldosterone release: A low sodium concentration increases aldosterone production, causing the kidneys to reabsorb more sodium, and vice versa.

Glucocorticoids (Corticosteroids)

CORTISOL (HYDROCORTISONE)

The principal glucocorticoid, cortisol, is produced in the zona fasciculata and is sometimes referred to as the "stress hormone." Like the catecholamines from the adrenal medulla, it helps the body respond during times of stress (for example, injury or emotional trauma). Cortisol is essential because of its effects on metabolism: It maintains the body's energy (glucose) supply and regulates fluid balance. Without it, the body could

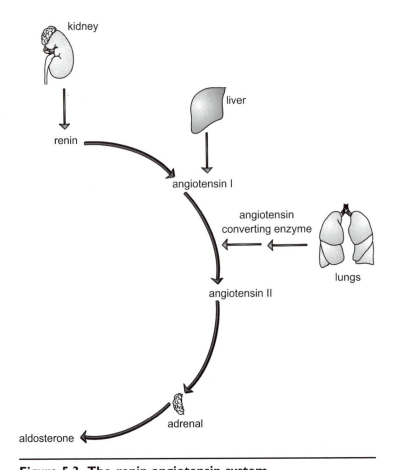

Figure 5.3. The renin-angiotensin system.

(1) Receptors in the kidney release renin into the blood in response to blood pressure changes; (2) in the liver, renin converts angiotensin into angiotensin I; (3) in the lungs, angiotensin I is converted into angiotensin II; (4) angiotensin II stimulates the adrenal gland to release aldosterone.

overreact to stress and disrupt the fragile homeostatic balance that it needs to stay alive.

The testes, erythrocytes, kidney medulla, and especially the brain rely on glucose as their sole energy source. Without it, they cannot function properly. After a meal, blood levels of glucose are typically high and the body is able to store whatever it does not immediately use. But glucose stores are not everlasting. Glycogen, the form of glucose stored in the liver, can run out within twenty-four hours after a meal. During periods of fasting, cortisol maintains blood glucose levels by affecting a number of metabolic processes:

GLUCOSE METABOLISM

In the muscles and other tissues, cortisol increases the breakdown of protein into amino acids. Those amino acids are used to produce additional glucose (via a metabolic pathway called **gluconeogenesis**) in the liver. Cortisol also conserves glucose for the brain and spinal cord by blocking the actions of insulin (which will be discussed later in this chapter)—inhibiting glucose absorption into other tissues.

FAT METABOLISM

Cortisol stimulates the release of fatty acids and **glycerol** from adipose tissue. Glycerol is used in gluconeogenesis, while fatty acids are made available for energy to other tissues to preserve glucose for the brain.

PROTEIN METABOLISM

Cortisol reduces protein reserves everywhere except in the liver. As proteins continue to be broken down in muscles and in other tissues, blood levels of amino acids rise. The additional amino acids are used for gluconeogenesis, glycogen formation, and protein synthesis in the liver.

Cortisol (as well as other, minor glucocorticoids) has a number of other important actions. It influences bone metabolism by lowering blood calcium levels. It does this by decreasing the absorption of calcium in the intestines and increasing the excretion of calcium by the kidneys. When blood calcium drops, the secretion of PTH increases (see Chapter 4). PTH pulls calcium from bones by stimulating the formation and activity of bone-dissolving cells called osteoclasts. Over time, high levels of glucocorticoids can lead to bone damage as more and more bone-building calcium is released into the blood.

Cortisol also maintains stable blood vessels and fluid volume. Without cortisol, blood pressure drops and kidney filtration declines as a result (see the Urinary System volume for more on kidney filtration). Cortisol stimulates the synthesis of erythropoietin by the kidneys, which increases red blood cell production. It also inhibits **fibroblasts,** the cells that form connective tissue, leaving the skin thinner and more easily bruised.

Cortisol serves as an anti-inflammatory and immunosuppressive agent. It inhibits the movement of inflammatory cells (monocytes and lymphocytes) to the site of injury, thereby reducing inflammation. It also suppresses cytokine synthesis and blocks antibody production, stifling the normal immune reaction. Doctors have harnessed cortisol's powerful immunosuppressive properties pharmaceutically to treat inflammatory and autoimmune diseases such as rheumatoid arthritis.

Cortisol aids in the maturation of the fetal lungs, liver, and gastrointestinal system. In males, glucocorticoids inhibit gonadotropin secretion, lowering testosterone levels. In females, glucocorticoids make the pituitary less

responsive to gonadotropin-releasing hormone (GnRH) from the hypothalamus, which reduces estrogen and progestin synthesis and inhibits ovulation.

In addition, cortisol, when present in excessively high or low quantities,
can alter behavior and emotion. People who produce too much of the hormone tend to be euphoric at first, but then fall into a state of depression.
High cortisol levels may also impair memory and concentration, decrease
sex drive, and cause insomnia. A cortisol deficiency usually leaves people
depressed, irritable, and apathetic. Low cortisol levels may also accentuate
an individual's sense of taste, hearing, and smell.

Cortisol release normally follows a circadian pattern, rising in the early
morning and declining throughout the day to reach its lowest levels while
an individual sleeps. But during a stressful situation, cortisol is released as
part of the stress response under the direction of the hypothalamic-pituitary
axis. When the body is faced with stress, the hypothalamus sends out neurotransmitters and corticotropin-releasing hormone (CRH) to signal the pituitary to release ACTH (see Figure 5.4). The more stress the individual
faces, the greater the CRH stimulation, and the more ACTH is subsequently
produced. Because of the need for a fast response, ACTH does not waste

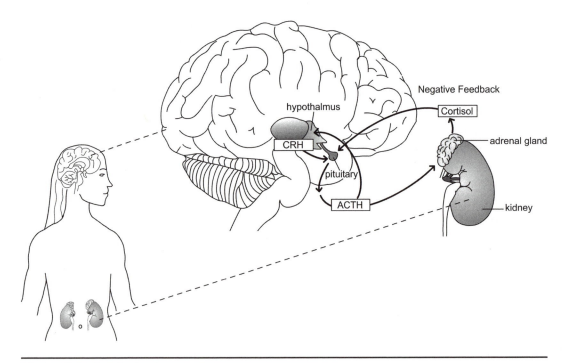

Figure 5.4. Cortisol release.
(1) The hypothalamus signals the pituitary to release ACTH, (2) the adrenal cortex releases cortisol into the blood, and (3) cortisol secretion is turned off by negative feedback upon the hypothalamus and pituitary.

any time. It can trigger cortisol secretion within minutes of reaching the adrenal cortex.

Upon stimulation by ACTH, the adrenal cortex produces cortisol from cholesterol and releases it into the blood. Cortisol circulates bound to blood plasma proteins, most commonly corticosteroid-binding globulin (transcortin). Nearly every cell in the body contains glucocorticoid receptors. Once it reaches its target cell, cortisol binds to the glucocorticoid receptor in the cytoplasm. The bound hormone-receptor complex travels to the nucleus, where it influences gene activity to stimulate or inhibit protein production.

Cortisol production is turned off by a classic negative feedback loop: Rising blood glucocorticoid levels inhibit CRH secretion by the hypothalamus, which turns off ACTH secretion by the anterior pituitary, stopping cortisol secretion by the adrenal cortex. The feedback can be fast, in which case ACTH secretion diminishes within minutes after blood cortisol levels rise; or the adrenal cortex may gradually become unresponsive to ACTH stimulation.

Gonadocorticoids

The adrenal cortex produces small quantities of the weak androgens **dehydroepiandrosterone (DHEA)** and **androstenedione** in the zona reticularis. Although they are not as potent as sex hormones from the gonads (testes and ovaries), adrenal androgens can be converted into stronger forms (testosterone and estrogen). These hormones influence the development of male and female characteristics.

Men and women produce both male and female sex hormones in the adrenal cortex. Testosterone converted from adrenal androstenedione is the primary source of male hormones in a woman's body, although in men it makes up just a tiny fraction of total testosterone. In both sexes, adrenal androgens influence sex drive as well as pubic and body hair growth. Even though women produce small amounts of male hormones, they usually don't exhibit masculine characteristics because they have enough circulating estrogen to overshadow the effects of androgens. After menopause, when estrogen production in the ovaries decreases, most of the estrogen in a woman's body comes from the conversion of androstenedione. But this small amount of estrogen is not enough to make up for the loss of ovarian estrogen, and postmenopausal women sometimes grow facial hair and display other masculine characteristics as a result.

Like cortisol, adrenal androgen release is controlled by the hypothalamic-pituitary axis. The hypothalamus and central nervous system release neurotransmitters and CRH, which trigger ACTH release by the anterior pituitary. ACTH acts upon the adrenal cortex to stimulate gonadocorticoid release. But unlike cortisol, gonadocorticoids do not inhibit hypothalamic release of CRH via negative feedback.

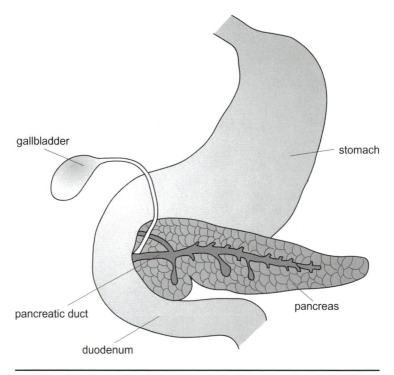

gallbladder

stomach

pancreatic duct

pancreas

duodenum

Figure 5.5. The pancreas.

THE ENDOCRINE PANCREAS

The pancreas (see Figure 5.5) functions as not one but two organs. The bulk of the organ is composed of exocrine tissue, which produces and sends special enzymes to the small intestine to aid in food digestion (see the Digestive System volume of this series). Scattered throughout the exocrine tissue are clusters of hundreds of thousands of endocrine cells called the islets of Langerhans. These cells produce the hormones insulin and glucagon, which regulate cellular energy usage and storage (see Figure 5.6). The islets also produce the hormones somatostatin and **pancreatic polypeptide.**

Energy Metabolism

Why are insulin and glucagon so crucial? Because the body needs energy to survive, and these two hormones regulate the distribution of energy to tissues. Energy enters the body in the form of food. As food passes through the mouth, esophagus, and stomach, enzymes break it down into tiny pieces. Once the partially digested food reaches the intestines, more enzymes go to work, breaking it down into molecules small enough to enter the bloodstream and be transported to cells. Starches are broken down into

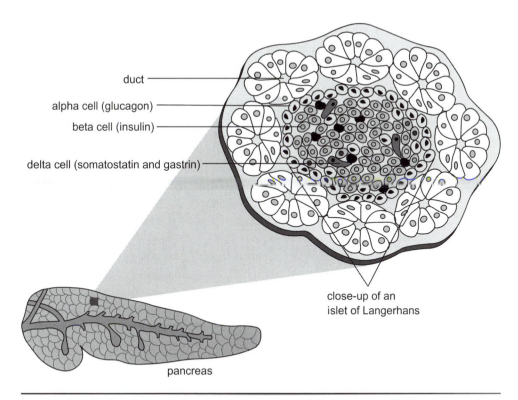

duct

alpha cell (glucagon)

beta cell (insulin)

delta cell (somatostatin and gastrin)

close-up of an
islet of Langerhans

pancreas

Figure 5.6. Cells of the endocrine pancreas and their respective hormones.

glucose (sugar), proteins are broken down into amino acids, and fats are broken down into fatty acids and glycerol.

Food metabolism occurs in two distinct phases: During the anabolic phase, which occurs after a meal, enzymes convert nutrients from food into substances the body can use. Blood levels of glucose, fatty acids, and amino acids rise. Because the body has more energy than it needs at the moment, it stores the excess for later. Glucose is stored as glycogen in the liver and muscles, fat is stored in adipose tissue, and amino acids are stored in muscle.

About four to six hours after a meal, the catabolic phase begins. Stored energy from the liver, muscles, and adipose tissue is mobilized to sustain the body until its next meal. The liver produces glucose from stored glycogen and by converting amino acids via gluconeogenesis. When the body has gone for some time without food, the liver converts free fatty acids into **ketone bodies.** The brain normally uses only glucose for energy, but it can use ketone bodies as a backup energy source when glucose supplies are low. Without this alternative energy source, the brain and nervous system would starve and suffer permanent damage.

The hormones insulin and glucagon from the endocrine pancreas regulate these stages of energy metabolism. Insulin primarily regulates the anabolic phase, while glucagon influences the catabolic stage.

Insulin

After a meal, the body converts carbohydrates from foods into simple sugars in the intestine. Glucose is carried to the tissues through the bloodstream. When blood glucose levels rise, the beta cells in the endocrine pancreas produce and release insulin. Insulin is formed from a larger, inactive molecule, called **proinsulin.** Before insulin is released into the bloodstream, the inactive molecule splits off.

Insulin levels rise 8–10 minutes after a meal, reaching their peak concentration 30–45 minutes after the meal. Nearly every cell in the body has insulin receptors. When insulin binds to its receptors on the cell surface, it triggers other receptors that help the cells take in glucose. The body uses and stores glucose in the liver, muscles, and adipose tissues. Without insulin, an individual could eat three meals a day and still starve to death because the cells would be unable to use the energy.

In the liver, insulin promotes glucose storage in the form of glycogen. It also inhibits the breakdown of glycogen and the production of glucose from other, noncarbohydrate sources (gluconeogenesis), and it decreases overall glucose release by the liver.

Insulin helps transport glucose to muscle cells and stimulates the incorporation of amino acids into protein, which is used to sustain and repair muscles. It promotes glycogen synthesis to replace glucose the muscles have used. Insulin also promotes glucose uptake in adipose tissue, promotes its conversion to fatty acids, and inhibits the release of stored fatty acids.

As insulin moves glucose into the tissues for energy use and storage, blood glucose levels fall. Between 90 and 120 minutes after a meal, blood glucose concentration returns to its original, premeal levels. To help the body maintain a constant blood glucose level, insulin and glucagon release are synchronized on an alternate schedule. When glucose concentrations in the blood rise during the anabolic phase, insulin is released. As insulin pulls glucose from the blood for tissue use and storage, blood glucose concentrations drop, stimulating glucagon release.

Insulin release may also be triggered by signals from the nervous system in response to external stimulation, for example the sight and/or smell of food. The gastrointestinal hormones cholecystokinin (CCK), secretin, gastrin, **gastric inhibitory peptide (GIP),** and acetylcholine are thought to play a role in this response, preparing the pancreas to release insulin. Insulin release is inhibited not only by low glucose levels but also by low levels of amino acids and fatty acids in the blood, as well as by the hormones somatostatin, epinephrine, and **leptin.**

Glucagon

Following a meal, insulin pulls glucose from the blood to be used and stored by cells. When several hours have passed without additional food being ingested, blood sugar is eventually depleted (a condition called hypoglycemia). The body still needs energy, much of which it gets from fatty acids until the next meal is available. But the brain, which cannot directly use fatty acids and other alternative energy sources, still relies on glucose. In response to dropping blood glucose levels, the alpha (α) cells of the endocrine pancreas begin to secrete the hormone glucagon from the large precursor molecule **proglucagon.** The same prohormone is found in cells of the gastrointestinal system, although it produces different secreted products.

Glucagon has the opposite effect of insulin. Whereas insulin lowers blood glucose levels by promoting glucose usage and storage, glucagon raises blood glucose levels (see Figure 5.7). It acts primarily upon the liver to in-

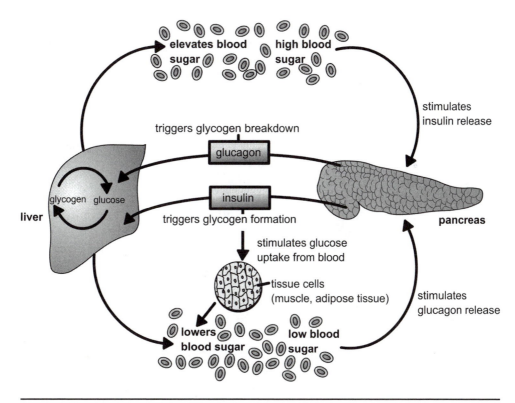

Figure 5.7. Actions of insulin and glucagon.

(1) Following a meal, insulin is released from the pancreas; (2) insulin stimulates the formation of glycogen (stored glucose) in the liver and stimulates glucose uptake by the muscles and adipose tissues; (3) as blood glucose levels drop, glucagon is released from the pancreas; and (4) glucagon promotes the breakdown of glycogen in the liver, raising blood sugar.

crease glucose output. When it binds to receptors on liver cells, glucagon activates the enzymes that break down stored glycogen (glycogenolysis) to release glucose and increases production of glucose from amino acid precursors (a process called gluconeogenesis). In adipose tissue, glucagon promotes the breakdown and release of fatty acids (**lipolysis**) into the blood, which are used by the cells for energy in the absence of glucose. By raising the level of fatty acids in the blood, glucagon indirectly prevents glucose uptake by the muscles and adipose tissue.

The main trigger for glucagon release is low blood sugar, but it may also be stimulated by other hormones, namely the catecholamines (in stressful situations); cholecystokinin, gastrin, and gastric inhibitory peptide (GIP) from the gastrointestinal system; and the glucocorticoids. Sympathetic nerve stimulation can also lead to glucagon release. Rising blood glucose levels, high circulating levels of fatty acids, as well as the hormones insulin and somatostatin inhibit glucagon secretion.

Insulin:Glucagon Ratio

Because the actions of insulin and glucagon are so closely interrelated to maintain a constant blood glucose level, the ratio of the two hormones is more important than the level of either hormone individually. When the ratio of insulin to glucagon is high, the tissues take in and store more glucose. When the ratio is low, nutrients are freed for use. A few hours after a meal, the insulin:glucagon ratio is at its normal rate of about 2:1, but it can drop to as low as 1:2 when the body has been fasting for a prolonged period, and it can rise to 10:1 or more after a carbohydrate-rich meal. A protein-rich meal will increase both insulin and glucagon production. Insulin increases the amount of amino acids pulled into muscles and stimulates protein production, while glucagon buffers the effects of insulin to prevent hypoglycemia.

Somatostatin

The hormone somatostatin is produced in the delta (δ) cells of the pancreatic islets as well as in the gastrointestinal tract and hypothalamus. Somatostatin is primarily an inhibitory agent. In the pancreas, it acts in a paracrine manner, suppressing production of insulin and glucagon. It also acts upon the gastrointestinal tract, inhibiting secretion of the hormones gastrin, secretin, and cholecystokinin; prolonging gastric emptying time; decreasing gastric acid and gastrin production; and slowing intestinal motility. Together, these actions reduce the rate of nutrient absorption. When secreted from the hypothalamus, somatostatin acts upon the pituitary to inhibit growth hormone secretion (see Chapter 3).

Somatostatin release is triggered by rising levels of glucose, fatty acids, and amino acids in the blood. Gastrointestinal hormones like secretin and cholecystokinin can also stimulate its release. Insulin inhibits somatostatin secretion.

Gastrin

Gastrin that is secreted by the stomach plays an important role in digestion. But scientists are still unsure as to what function gastrin has when it is secreted by the δ cells of the endocrine pancreas.

Pancreatic Polypeptide

Pancreatic polypeptide (PP) is produced by the F cells of the pancreas. It is secreted following a protein-rich meal or in response to hypoglycemia or strenuous exercise. PP inhibits gallbladder contractions and pancreatic enzyme secretion. High levels of free fatty acids in the blood inhibit its release.

GASTROINTESTINAL HORMONES

The intestinal tract not only digests food and absorbs nutrients; it also produces and secretes a number of hormones that aid in the digestive process. Gastrointestinal hormones, which are primarily peptides, are produced in specialized endocrine cells in the stomach and small intestine (see Figure 5.8), as well as in neurons scattered throughout the gastrointestinal tract. Most gastrointestinal hormones either act upon nearby cells (paracrine delivery) or act as neurotransmitters within neurons (neurocrine delivery). Endocrine and neural cells in the intestinal tract are referred to collectively as the enteric endocrine system.

The central nervous system and gastrointestinal tract are linked by pathways known as the brain-gut axis. Neurotransmitters located in both the brain and the gut regulate and coordinate such functions as satiety, nutrient absorption, gut motility, and intestinal blood flow. Any disruption of this system is believed to result in gastrointestinal disorders such as irritable bowel syndrome (IBS).

Gastrointestinal Hormones

GASTRIN

This hormone, produced by specialized cells in the stomach, regulates stomach acid secretion. When partially digested proteins, peptides, and amino acids are present in the stomach, gastrin stimulates the release of gastric acid and the digestive enzyme, pepsin, into the stomach cavity to aid digestion. Beer, wine, and coffee can also stimulate gastrin release. As the stomach becomes more acidic, gastrin production declines.

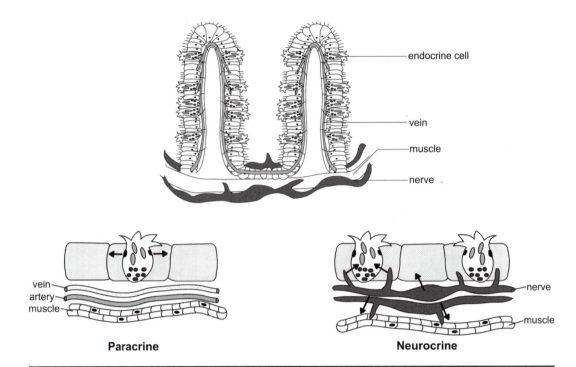

Figure 5.8. Endocrine cells in the wall of the small intestine.

CHOLECYSTOKININ (CCK)

Food entering the small intestine must be broken down into smaller molecules (such as amino acids and fatty acids) in order to be absorbed. When partially digested proteins and fats enter the duodenum (first portion) of the intestine, cells in that region secrete the peptide hormone CCK. This hormone triggers the release of pancreatic enzymes and stimulates gallbladder contractions to release stored bile. Pancreatic enzymes and bile are sent to the intestines, where they aid in digestion. As proteins and fats are digested and absorbed, a drop in their levels shuts off CCK release. CCK is also released as a neurotransmitter by the central nervous system. Scientists believe that it may help regulate food intake by signaling the feeling of satiety.

SOMATOSTATIN

Somatostatin was first isolated in the hypothalamus for its role as a growth hormone inhibitor. Later, it was also found in cells within the endocrine pancreas and gastrointestinal tract. In the gut, somatostatin inhibits the release of various hormones, including gastrin, CCK, secretin, and vasoactive intestinal polypeptide (VIP). It also inhibits secretion of gastric acid, pancreatic enzymes, and bicarbonate. In larger doses, somatostatin inhibits smooth muscle contractions and blood flow in the intestine. Unlike most gut peptides, it can act via endocrine, paracrine, neurocrine, and autocrine delivery. Somatostatin secretion is stimulated by the ingestion of a high-protein, high-fat meal. It inhibits its own release in an autocrine fashion.

SECRETIN

The stomach regularly secretes acid, which could potentially burn and damage the small intestine. When acid enters the duodenum, cells lining the region release secretin. This hormone stimulates the pancreas to release acid-neutralizing bicarbonate and water. As acid in the intestine is neutralized, the rising pH level shuts off secretin release. Somatostatin also inhibits secretin production.

GASTRIC INHIBITORY PEPTIDE (GIP)

Gastric inhibitory peptide (GIP), also called glucose-dependent insulinotropic peptide, is part of the secretin family of hormones, and like secretin it is released by cells in the duodenum. It blocks gastrin and gastric acid secretion into the stomach and inhibits gut motility. Following a meal, GIP responds to increased glucose levels, enhancing the insulin response to glucose.

VASOACTIVE INTESTINAL POLYPEPTIDE (VIP)

This peptide is found throughout the body but is secreted in greatest concentration by cells in the intestinal tract and nervous system. VIP increases secretion of water and electrolytes by the intestine, increases blood flow within the gut, and inhibits gastric acid secretion.

GHRELIN

This peptide hormone, secreted by epithelial cells in the stomach, stimulates growth hormone secretion from the pituitary gland. In the gastrointestinal system, ghrelin stimulates the sensation of hunger by communicating the body's energy needs to the brain. Ghrelin levels in the blood are highest during the fasting state several hours after a meal, and lowest just after food consumption. Scientists are investigating ghrelin's role in obesity, in the hope of one day discovering more effective weight control methods.

MOTILIN

As its name suggests, **motilin** controls movement (smooth muscle contractions) in the gut. In between meals, cells in the duodenum secrete small bursts of motilin into the blood at regular intervals. Motilin contracts and releases smooth muscles in the intestine wall to clean undigested materials from the intestine.

SUBSTANCE P

The neuropeptide **substance P,** found in both the brain and gastrointestinal system, stimulates smooth muscle contractions and epithelial cell growth. It may also be involved in inflammatory conditions of the gut. In the brain, substance P has been linked to both the pain and pleasure responses. It is released from enteric neurons in response to central nervous system stimulation, serotonin, and CCK, and it is inhibited by somatostatin.

6

Endocrine Functions of the Sex Glands

The ovaries and testes are necessary for reproduction: Without them, the human species could no longer reproduce. Along with their central functions—producing and sustaining the eggs and sperm—the sex glands have a number of other crucial tasks. As part of the endocrine system, they produce and secrete the sex hormones, which are involved in sexual maturation, differentiation, and function, as well as metabolism and bone growth.

THE OVARIES

The ovaries (see Figure 6.1) are the female reproductive organs. Inside the ovaries, the eggs, or ova, develop and are nourished until they are ready to be released into the fallopian tubes. The eggs form and develop by a process known as **oogenesis.** In the developing female embryo are primordial germ cells, from which the fundamental egg cells develop. These cells remain dormant until a girl reaches puberty, then develop into the mature eggs, which are released one at a time in a process known as ovulation. Like the adrenal gland, each ovary is constructed of an outer cortex and an inner medulla. The cortex contains tiny sacs called follicles. Ovarian follicles (see Figure 6.2) consist of two types of cells: theca and granulosa. Nestled inside each follicle is an immature egg, surrounded by a layer of granulosa cells. A woman is born with every egg (about a million of them) that she will ever possess. Once these eggs are used up, she will no longer be able to conceive.

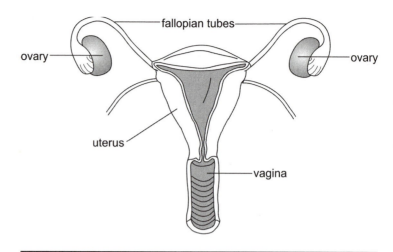

Figure 6.1. The ovaries.

The ovarian follicles normally remain in an inactive state. These so-called primordial follicles lie in wait for hormonal stimulation that will help them mature and prepare them for possible fertilization. During each menstrual cycle, follicle-stimulating hormone (FSH) from the pituitary stimulates a few of these eggs. Typically, only one egg completes the ovulation process.

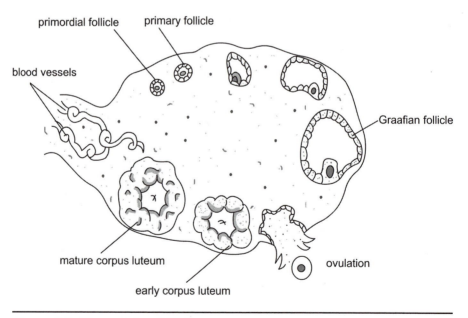

Figure 6.2. The ovarian follicle.

Menstrual Cycle Phases

The monthly menstrual cycle is a complex interaction involving the hypothalamus, anterior pituitary, and ovaries. On average, the cycle takes twenty-eight days to complete. It is divided into four phases: follicular, ovulatory, luteal, and menstruation.

FOLLICULAR PHASE

During this phase, which lasts 10–16 days, gonadotropin-releasing hormone (GnRH) from the hypothalamus triggers the release of follicle-stimulating hormone (FSH) and luteinizing hormone (LH) from the anterior pituitary. FSH and estrogen stimulate the development of between six and twelve primary follicles. LH causes the follicles to produce estrogen. Between days 5 and 7 of the cycle, one of the follicles ripens (in the case of a multiple pregnancy, more than one follicle has ripened) and becomes ready for ovulation. As the dominant follicle develops, estrogen levels rise, inhibiting FSH secretion. Without FSH stimulation, the other follicles begin to wither away.

OVULATORY PHASE

At the end of the follicular cycle, estrogen levels peak. Estrogen normally suppresses gonadotropin production, but in this case, rising estrogen levels trigger a surge in LH and FSH (usually around day 13 of the cycle). The hormonal surge lasts between twenty-four and thirty-six hours, at the end of which the dominant follicle ruptures and releases its egg from the ovary. This is called ovulation.

LUTEAL PHASE

Following ovulation, the erupted follicle transforms into a body called the corpus luteum. The corpus luteum begins to secrete estrogen and progesterone under the influence of luteinizing hormone from the pituitary. Estrogen and progesterone prepare the uterus for implantation and are necessary to maintain a pregnancy. If the egg is fertilized, the corpus luteum remains intact and continues to secrete estrogen and progesterone throughout the first trimester of pregnancy. The luteal phase lasts for about fourteen days.

MENSTRUAL PHASE

If the egg is not fertilized, production of progesterone and estrogen in the corpus luteum diminishes. The uterine lining, which has become rich with blood vessels to nourish the growing embryo, is no longer needed. Small arteries in the lining constrict, cutting off oxygen and nutrients. The cells die and slough off (known as menstruation). The corpus luteum undergoes a process known as **luteolysis.** It degenerates, becomes unable to produce hormones, and is finally replaced by scar tissue. As steroid production by the corpus luteum decreases, FSH secretion increases, stimulating the development of new follicles and initiating a new menstrual cycle. Menstruation lasts for four to five days.

Ovarian Hormones

In addition to providing a site for egg development, the ovaries produce several steroid and peptide hormones. Steroid hormones (estrogens, progesterone, and androgens) are produced in the follicular cells. Like all steroid hormones, they are produced from cholesterol, which is both present in the ovaries and transported to the ovaries in the form of low-density lipoproteins (LDLs). Peptide hormones (relaxin, inhibin, oxytocin, and vasopressin) are produced in the follicular cells and within the corpus luteum.

ESTROGENS

The primary female sex hormones produced in the ovaries, estrogens play a role in the development of sexual characteristics and help regulate the reproductive cycle. The three estrogens—**estradiol, estriol,** and **estrone**—are produced in the thecal and granulosa cells in the developing follicles.

Estradiol is the most powerful—and most plentiful—of the three estrogens. It is made from the androgens testosterone and androstenedione, which are produced in the thecal cells under the influence of luteinizing hormone. In the granulosa cells, follicle-stimulating hormone helps convert these androgens into estradiol (and estrone). At the onset of puberty, estradiol influences maturation of the reproductive organs (uterus, fallopian tubes, cervix, and vagina) and redistributes fat to the hips, buttocks, thighs, and breasts to produce a more feminine, curvy shape.

Estradiol also influences the menstrual cycle by stimulating and inhibiting the release of LH and FSH. Normally, estradiol acts upon the hypothalamus to inhibit GnRH secretion, which prevents the release of LH from the anterior pituitary. But at the end of the follicular phase of the menstrual cycle, rising estradiol concentrations trigger the surge of LH that initiates ovulation.

Most of the estriol in a woman's body is produced not in her ovaries but in her liver, where it is converted from estrone and (by a more indirect route) estradiol. This relatively weak hormone may actually act as a partial agonist–partial antagonist by blocking receptors that would otherwise be occupied by the stronger estrogen, estradiol. During pregnancy, estriol is secreted in large quantities by the placenta. Doctors test a mother's urine for this hormone to assess the viability of her pregnancy.

Estrone, the weakest of the three estrogens, is primarily converted from estradiol or androstenedione (from the adrenal cortex). It is similar in function, although not as potent, as estradiol. Following menopause, estrone production increases due to increased conversion of androstenedione.

When the estrogens are secreted into the bloodstream, they travel bound to proteins—mainly albumin and **sex hormone–binding globulin (SHBG).** Once they arrive at their target cells, estrogens bind to an intracellular protein, which carries them to the nucleus. There, they influence protein synthesis.

In addition to the roles outlined in this chapter, estrogens have several other functions:

Bone growth: During puberty, estrogens promote closure of the **epiphyses** in the long bones, which stops bone growth. They also oppose the actions of parathyroid hormone (PTH) by increasing calcium resorption by the bones.

Metabolism: Estrogens influence metabolism by slightly impairing the body's ability to metabolize glucose (in opposition to insulin) and by increasing the levels of triglycerides (fats) and decreasing the levels of cholesterol and low-density lipoproteins (LDLs) in the blood.

Gastrointestinal system: Estrogens reduce bowel motility, which can affect nutrient absorption in the intestine.

Kidneys: Estradiol increases sodium reabsorption in the kidney tubules. The resulting rise in blood sodium levels can cause some women to retain water.

PROGESTINS

Progestins are primarily designed to maintain and support a pregnancy. Synthetic versions can also prevent a pregnancy (see "Harnessing the Power of Hormones to Prevent a Pregnancy"). The most significant progestin is progesterone. It is produced by the corpus luteum after the egg has been released from the follicle. If the egg is fertilized, the corpus luteum continues to produce progesterone for the first trimester of pregnancy until the placenta takes over production. Progesterone prepares the mother's body for pregnancy by thickening the uterine lining to nourish the growing embryo. It then maintains the viability of the pregnancy by stopping additional follicles from becoming mature and by preventing uterine contractions.

Progesterone travels through the blood bound to corticosteroid-binding globulin (CBG) and albumin. Its other actions are to stimulate breast growth and development (along with estrogen); influence carbohydrate, protein, and fat metabolism; and decrease the body's responsiveness to insulin (as sometimes occurs during pregnancy).

OTHER OVARIAN HORMONES

The peptide hormone relaxin is produced by the corpus luteum during pregnancy in response to human chorionic gonadotropin (hCG). It relaxes the pelvic ligaments and softens the cervix to facilitate childbirth. The peptide hormone inhibin is produced in the granulosa cells of the ovaries (as well as in the Sertoli cells in the male testes) and inhibits the secretion of follicle-stimulating hormone (FSH) from the anterior pituitary. Activin is a

Harnessing the Power of Hormones to Prevent a Pregnancy

The sex hormones estrogen and progesterone maintain the reproductive cycle as well as establish and support a pregnancy. But in the 1950s, scientists discovered that synthetic versions of these hormones could also be used to prevent pregnancies. In 1960, the U.S. Food and Drug Administration approved the first oral contraceptives, which contained a combination of estrogen and progestin. These combination pills, which prevent a pregnancy by stopping the release of eggs from the ovaries, are still in use today. Other birth control pills containing progesterone only (called the "minipill") thicken the cervical mucus to prevent the sperm from reaching the egg.

Since birth control pills were first introduced, research has come up with a whole new range of delivery methods. Depo-Provera is similar to the minipill in that it contains progestin only, but it is injected once every three months rather than taken in pill form once a day. Norplant consists of tiny rods filled with progestin, which a doctor implants under the skin of a woman's arm. The implants provide protection for up to five years. Ortho Evra, which was approved in 2001, releases estrogen and progestin into the bloodstream through a patch affixed to the skin on the lower abdomen, buttocks, or upper body. And NuvaRing releases estrogen and progestin directly into the vagina via an inserted ring.

protein associated with inhibin that stimulates FSH secretion. Furthermore, the ovaries produce three androgens: the weak male hormone androstenedione as well as small quantities of the more potent hormones **dihydrotestosterone (DHT)** and testosterone. Androgens play a role in pubic and body hair growth during puberty, and they work with estrogens to stimulate bone growth.

THE TESTES

The testes (see Figure 6.3), like the female ovaries, serve both reproductive and endocrine functions. They are made up of a network of tubules (seminiferous tubules) that produce and carry sperm, interspersed with cells in which androgens (male hormones) are produced. The testes contain two types of cells:

> *Sertoli cells*: These cells, which line the seminiferous tubules, surround and nourish the germ cells from which sperm develop, and they facilitate the journey of sperm out of the testes. Sertoli cells also produce androgen-binding protein, which maintains high levels of androgens in the testes and seminal fluids, as well as many peptides (inhibin, activin, and follistatin) that regulate testicular function.

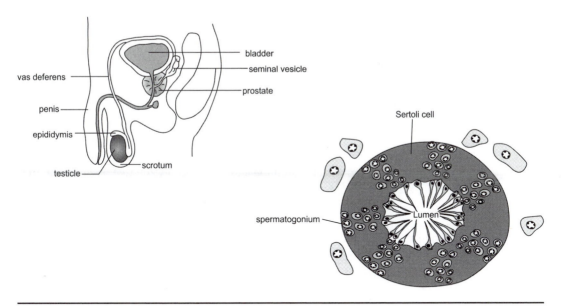

Figure 6.3. The testes.

Leydig cells: In between the seminiferous tubules are the cells in which testosterone, the primary androgen, is produced.

Sperm Production

Production of sperm in the Sertoli cells depends upon stimulation by follicle-stimulating hormone (FSH) from the anterior pituitary and by testosterone. When FSH binds to androgen receptors on Sertoli cells, it stimulates the production of androgen-binding protein. This protein keeps levels of testosterone in the seminiferous tubules high. FSH stimulates sperm production and maturation. As androgen levels rise, Sertoli cells begin to secrete inhibin, which inhibits FSH release from the pituitary.

Testicular Hormones

The testes produce a number of hormones, including testosterone, dihydrotestosterone, androstenedione, and estradiol. Testosterone is by far the most plentiful, and most important, of these hormones.

Like other steroid hormones, testosterone is synthesized from cholesterol. Its release is initiated by gonadotropin-releasing hormone (GnRH) from the hypothalamus, which stimulates the release of luteinizing hormone from the pituitary. LH stimulates the Leydig cells to produce and secrete testosterone. In a classic negative-feedback loop, elevated testosterone levels in the blood inhibit secretion of LH by acting on the hypothalamus and the anterior pituitary, both of which contain androgen receptors.

Testosterone travels through the blood bound to carrier proteins—typically either sex hormone–binding globulin (SHBG) or albumin. When it reaches its target cell, it binds to the androgen receptor.

Testosterone can work on its own, but it is more commonly converted by enzymes into the more potent dihydrotestosterone (DHT). The testes also produce small amounts of DHT.

DHT and testosterone serve many important functions:

Sexual differentiation: In the fetus, androgens are responsible for differentiation of the external male genitalia and internal reproductive organs (including the penis, epidydymis, vas deferens, seminal vesicles, and scrotum).

Puberty: At puberty, a testosterone surge transforms a boy into a man. It causes the larynx to enlarge, deepening the voice; makes the penis and testes grow larger; leads to hair growth on the face, body, and pubic area; and increases height and muscle mass.

Reproduction: Testosterone directs the development of the seminiferous tubules, promotes and maintains sperm production (spermatogenesis), and increases a man's sex drive.

The process of oogenesis (egg development) and spermatogenesis (sperm production) are covered in the Reproductive System volume of this series.

7

Endocrine System Development

Beginning in the womb and continuing throughout an individual's life, the endocrine system directs most of the body's physiological processes. Hormones oversee and regulate all of life's most important changes: the emergence and development of the fetus in the womb, the transformation of the child to an adult at puberty, and the physical and emotional change of life that accompanies the aging process.

PREGNANCY

During pregnancy, a woman's body undergoes a series of changes, many of which involve the endocrine system. Just to achieve a successful pregnancy requires the efforts of several hormones, including luteinizing hormone (LH), follicle-stimulating hormone (FSH), estrogen, and testosterone (see Table 7.1). Both sperm and eggs depend upon hormones for their existence and to help them survive the journey to fertilization and implantation (see Chapter 6, as well as the Reproductive System volume of this series, for more on this topic). Once a pregnancy has been established, hormones nourish and protect the growing fetus, and ease the baby's tumultuous journey into the world.

Almost immediately after fertilization occurs, the embryo begins releasing a protein called human chorionic gonadotropin (hCG), which is the primary indicator by which pregnancy tests detect conception. Under the direction of hCG, the corpus luteum (which emerges out of the ovarian follicle after the egg is released) secretes the hormones estradiol and progesterone, and continues to secrete them throughout the first trimester of pregnancy. These

TABLE 7.1 Reproductive Hormones

Hormone	Function
Luteinizing hormone (LH) and follicle-stimulating hormone	Help establish a pregnancy.
Progesterone	Protects the pregnancy by preventing uterine contractions and ovulation, and by providing a hospitable environment for the fetus.
Estrogen	Improves uterine blood flow and stimulates uterine growth.
Human chorionic gonadotropin (hCG)	Triggers estrogen and progesterone release from the corpus luteum.
Human placental lactogen (hPL)	Provides glucose and fatty acids to nourish the growing fetus.
Relaxin	Softens mother's pelvis to ease childbirth.

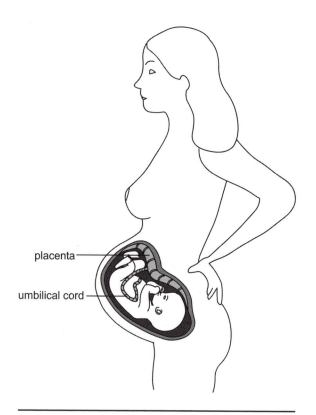

placenta

umbilical cord

Figure 7.1. The placenta.

hormones help keep the pregnancy viable. As the pregnancy progresses, the placenta begins to take over estrogen and progesterone production.

The placenta (see Figure 7.1) protects and nourishes the growing fetus and serves as the fetal respiratory and excretory organs. It also acts as an endocrine organ, secreting several steroid and protein hormones and providing precursor molecules for the fetal endocrine system until it is developed enough to produce and secrete its own hormones. The placenta also stimulates maternal hormone production. The coordination between the developing fetal endocrine system and the placenta is referred to as the fetal-placental unit.

Steroid Hormones

PROGESTERONE

In the first few weeks of pregnancy the corpus luteum supplies most of the progesterone for the developing fetus, but the placenta takes over by the beginning of the second trimester. Because progesterone production requires cholesterol (see Chapter 1 on steroid hormone production), which the placenta is unable to synthesize on its own in large quantities, it is forced to rely on cholesterol from the mother that arrives in the form of low-density lipoproteins (LDLs).

Progesterone protects the pregnancy in several ways. It inhibits uterine contractions to prevent a premature delivery (near the end of the pregnancy, progesterone levels decrease, allowing the uterus to contract and expel the baby); provides a hospitable environment for the fetus in utero; inhibits follicle-stimulating hormone (FSH) and luteinizing hormone (LH) secretion, preventing ovulation (see Chapter 6); and helps prevent an immune response that might cause the mother's body to reject her unborn baby. Finally, it contributes to the production of a thick, gelatinous substance called a mucous plug. The mucous plug stops up the cervical canal, preventing bacteria from entering the uterus and reaching the embryo. This plug is typically expelled a few days before the mother gives birth.

ESTROGEN

As is the case with progesterone, the mother's ovaries provide most of the estrogen that is needed during the first trimester. Later, the placenta takes over estrogen production. The level of circulating estrogens can rise several hundredfold during the course of a pregnancy. Estrogen improves uterine blood flow and stimulates uterine growth to accommodate the growing fetus.

The primary estrogen produced by the placenta is estriol. Urine and blood tests of this hormone are used to assess fetal health in high-risk pregnancies. Estrone and estradiol are also produced in smaller concentrations. Because the placenta lacks the necessary enzymes to synthesize estrogens from pregnenolone and to synthesize estriol from estradiol, producing these hormones requires a complex coordination involving the placenta, fetal liver, and adrenal glands. The maternal and fetal adrenal glands convert cholesterol into the steroid hormone dehydroepiandrosterone (DHEA). In the liver, DHEA is converted into dehydroepiandrosterone sulfate (DHEA-S). The placenta then metabolizes DHEA sulfate into estriol.

Protein Hormones

HUMAN CHORIONIC GONADOTROPIN (hCG)

Within eight to ten days after the embryo has attached itself to the uterine lining, the placenta begins producing the protein hormone hCG. Pregnancy tests can detect the presence of hCG in the mother's blood and urine, and thus confirm that conception has occurred, within ten days of fertilization. During the first few weeks of pregnancy, hCG levels rise rapidly, doubling every 36–48 hours before reaching their peak at the end of the first trimester. hCG is similar in structure to luteinizing hormone (LH). Like LH, hCG triggers the release of the steroid hormones estrogen and progesterone from the corpus luteum.

HUMAN PLACENTAL LACTOGEN (hPL)

Human placental lactogen (hPL), also called human chorionic somatomammotropin, is similar both in structure and in function to both growth hormone and prolactin. hPL levels rise steadily throughout the pregnancy and are first detectable at about 4–5 weeks' gestation. When this hormone enters the mother's bloodstream, it alters the way her body metabolizes glucose and fatty acids. hPL increases the mother's blood glucose levels while inhibiting glucose uptake into her tissues, freeing additional energy to nourish her growing baby. hPL also releases free fatty acids to be used by both the mother and her baby. Because hPL blocks the actions of insulin (known as "insulin resistance"), it can contribute to gestational diabetes.

RELAXIN

The hormone **relaxin** is believed to act in conjunction with progesterone to help maintain the pregnancy. Relaxin also softens the mother's pelvic ligaments to ease childbirth.

The placenta also produces several hormones that are structurally and functionally similar to those produced by the hypothalamus (placental gonadotropin-releasing hormone [GnRH], corticotrophin-releasing hormone [CRH], and thyrotropin-releasing hormone [TRH]) and pituitary (placental growth hormone [GH] and adrenocorticotropic hormone [ACTH]).

Pregnancy and the Maternal Endocrine System

Placental hormones not only circulate in the unborn baby's bloodstream—they also travel through the mother's system. These hormones have the power to influence nearly every endocrine gland in the mother's body.

In addition to hormone-induced changes, the mother's endocrine system temporarily evolves in order to nourish and protect her growing baby (see Figure 7.2). The anterior pituitary grows, often doubling in size. FSH and LH levels stay low during pregnancy because they are less responsive to gonadotropin-releasing hormone (GnRH) from the hypothalamus (due to rising estrogen and progesterone levels). Prolactin secretion rises, peaking at delivery to stimulate milk production. The posterior pituitary releases oxytocin, which causes the uterus to contract during delivery and causes the breasts to release milk following delivery; and vasopressin, which constricts the mother's blood vessels and helps her retain more fluid.

The thyroid gland increases slightly as the demand for thyroid hormones increases. Blood concentrations of T4-binding globulin (TBG) rise. TBG is one of several proteins that transport thyroid hormones in the blood. When

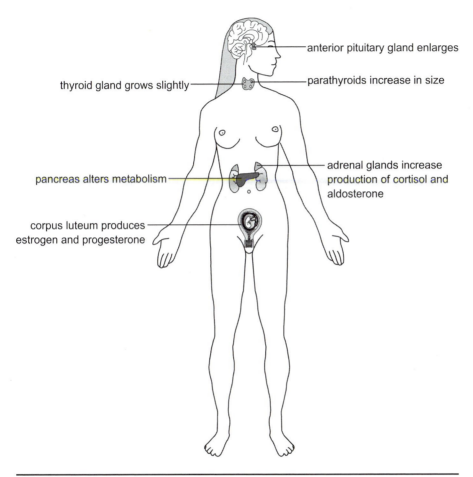

anterior pituitary gland enlarges

parathyroids increase in size

thyroid gland grows slightly

pancreas alters metabolism

adrenal glands increase
production of cortisol and
aldosterone

corpus luteum produces
estrogen and progesterone

**Figure 7.2. Changes to the maternal endocrine system during
pregnancy.**

its levels rise, pituitary release of thyroid-stimulating hormone (TSH) also
increases, resulting in greater secretion of the thyroid hormones T_4 and T_3.
The increased thyroid hormone production taxes maternal iodide stores. If
the mother does not get enough iodine in her diet, hormone synthesis may
be affected, and she may suffer from an iodide deficiency. Maternal thyroid
hormones are crucial for fetal brain development and are the only source of
thyroid hormones until the fetal thyroid becomes fully operational. If a child
does not receive enough thyroid hormones in utero and in the first two years
of life, he or she may suffer from mental retardation or other cognitive impairment (see Chapter 11).

The parathyroid gland increases in size and produces more parathyroid
hormone to meet the fetal need for vitamin D and calcium. Also, the pan-

creas alters its function to meet the changing nutritional needs of both mother and baby. In early pregnancy, insulin levels stay steady or may even drop in response to falling blood glucose levels in the mother. But as the pregnancy progresses, insulin levels rise. Insulin is not transported across the placenta to the fetus, but its effects on maternal metabolism indirectly reach the fetus. The new pregnancy-altered metabolism preserves glucose while meeting the mother's energy needs through fatty acid metabolism. As insulin levels rise, the mother becomes more resistant to its effects, making her susceptible to gestational diabetes.

The adrenal cortex produces more cortisol, which influences protein and carbohydrate metabolism. Aldosterone levels also rise, reaching their peak in the middle of the pregnancy. Aldosterone influences sodium and potassium absorption (see Chapter 5).

Embryonic Endocrine System Development

In its earliest stages of development, the fetal endocrine system relies on precursor hormones secreted by the mother and placenta. But as the fetus matures, its endocrine system develops to the point where it can produce and secrete its own hormones. The endocrine system is one of the first systems to develop in the fetus. Endocrine organs develop from specialized cells in the embryo. The thyroid and pancreas develop from cells in the embryonic digestive system; the parathyroid and adrenal medulla arise from nervous system cells; and the ovaries, testes, and adrenal cortex emerge from an area called the urogenital ridge. The pituitary gland is derived from cells originating in both the nervous and digestive systems.

HYPOTHALAMUS

The hypothalamus begins to emerge during the first few weeks of embryonic life and is advanced enough to begin releasing its hormones (including gonadotropin-releasing hormone [GnRH], thyrotropin-releasing hormone [TRH], and corticotropin-releasing hormone [CRH]) by about eight weeks.

ANTERIOR PITUITARY

The anterior and posterior parts of the pituitary gland develop individually (see Figure 7.3). The anterior pituitary emerges out of cells lining an outpouching on the roof of the mouth (Rathke's pouch). The cells differentiate into five individual types, delineated by the type of hormone that each secretes (gonadotroph, thyrotroph, corticotroph, somatotroph, and lactotroph). By seven weeks, these early pituitary cells can secrete follicle-stimulating hormone (FSH), luteinizing hormone (LH), growth hormone, prolactin, and corticotropin (ACTH). At the end of the first trimester, the anterior pituitary is producing all of its hormones, as is the hypothalamus, although the hypophyseal portal system connecting the two bodies is not complete until late in the second trimester. During the third trimester, the feedback mechanisms that stimulate and inhibit pituitary hormone release become more sophisticated.

POSTERIOR PITUITARY

The posterior pituitary develops out of a downward extension at the base of the developing brain. It remains attached to the brain by a stalk (the hypophyseal, or pituitary, stalk) of nerve fibers. Eventually, the anterior and posterior pituitary merge, but they continue to have distinct functions. By midpregnancy, the posterior pituitary begins to secrete oxytocin and antidiuretic hormone (ADH). The intermediate pituitary virtually disappears after birth—only interspersed cells remain in the adult.

THYROID

The thyroid is one of the first organs to emerge, and its hormones play a central role in fetal brain and central nervous system development. The gland develops from an outpouching on the floor of the pharynx, and it grows

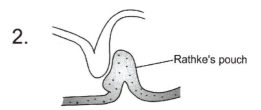

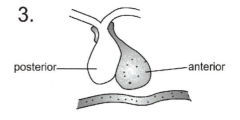

Figure 7.3. Development of the fetal pituitary.

downward in front of the trachea in the neck. Finally, two lobes are revealed, connected by a thin strip of tissue called the isthmus. By the end of the second trimester, the thyroid is able to trap iodine. Thyrotropin-releasing hormone (TRH) and thyroid-stimulating hormone (TSH) are present. The thyroid begins to secrete its hormones by about eighteen to twenty weeks. Soon, the feedback mechanisms governing thyroid hormone release are in place. Few thyroid hormones cross the placenta from the mother, but the fetus does receive maternal TRH and iodide, which it needs to produce thyroid hormones. Because of this transfer, the fetus may be susceptible if the mother suffers from thyroid dysfunction.

PARATHYROIDS

The parathyroids develop out of branchial pouches—grooves that re-
semble the gills of a fish—in the neck of the embryo. They are able to produce parathyroid hormone (PTH) by the end of the first trimester, but levels
remain low until birth. The fetus gets the calcium it needs for bone growth
from its mother. As maternal PTH levels rise, her blood calcium increases.
That additional calcium crosses the placenta to the developing fetus. Fetal
calcium levels rise consistently throughout the course of the pregnancy.

ADRENAL

The fetal adrenal gland (see Figure 7.4) is derived from two distinct tissue types. The adrenal cortex develops from the mesoderm layer, one of the
three layers in the developing embryo from which every cell, tissue, and
organ in the body emerge. By twenty weeks' gestation, the fetal adrenal gland
is larger than the adult gland (relative to body size). Most of the gland—80
percent—is made up of a fetal zone. The majority of steroid hormones produced in the fetal adrenal gland originate in this fetal zone. The outer region is called the definitive zone, which will later divide into the three layers
of the adult adrenal cortex (glomerulosa, fasciculata, and reticularis). The
definitive zone begins growing rapidly in the third trimester and takes over
hormone production postnatally. Near term, the adrenal gland secretes large
quantities of steroid hormones—up to five times that of an adult at rest.

The adrenal medulla arises from the neuroectoderm layer in the embryo.
The medulla is formed by about ten weeks' gestation, but unlike the cortex,
it is still relatively immature at birth. The cortex grows around and surrounds the medulla. Both epinephrine and norepinephrine are discernible
by the latter part of the first trimester. The adrenal gland remains large after
birth, but within a few months, the fetal zone regresses and the gland
shrinks to its normal size.

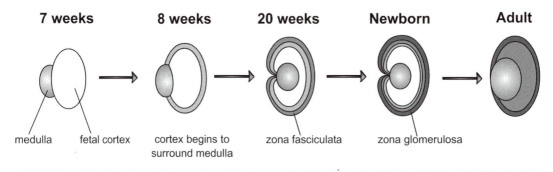

Figure 7.4. Development of the fetal adrenal gland.

PANCREAS

The pancreas appears during the fourth week of embryonic development. The first cells to emerge are the alpha cells, which produce glucagon, and the delta cells, which produce somatostatin. Insulin first appears before the beta cells have fully developed; however, it probably has less effect on fetal blood glucose levels than does glucose transferred from the mother across the placenta.

SEX GLANDS

The sex of the fetus is determined by chromosomes donated by the egg and sperm. Humans have twenty-three pairs of chromosomes: twenty-two autosomes and one pair of sex chromosomes. The mother provides the X sex chromosome, and the father provides either an X or a Y. If the embryo contains the chromosome pair XX, it will become a female; if it contains the chromosome pair XY, it will develop into a male (the effects of chromosomal abnormalities are covered in the Reproductive volume of this series). For the first seven or eight weeks of development, the male and female embryos look identical and have the potential to develop the genitalia of either sex. What determines sexual differentiation is a gene family in a region on the Y chromosome called SRY (sex-determining region). SRY initiates a series of chemical reactions that cause the primitive gonad to differentiate into the male testes. In the absence of a Y chromosome, the gonads develop into the female ovaries.

At the fifth or sixth week of development, an undifferentiated gonadal ridge develops. Out of this ridge form the Wolffian ducts, which serve as excretory ducts for the developing kidneys, and from which the male gonads will eventually develop. Also present are the Müllerian ducts, from which the female genitalia will develop.

In embryos that contain a Y chromosome and are thus male, SRY causes the undifferentiated gonad to differentiate into the male testes. The Sertoli cells (see Chapter 6) of the testes develop at about eight weeks' gestation and begin producing a peptide hormone called Müllerian-inhibiting substance (MIS). The presence of MIS and androgens causes the Müllerian ducts to disappear. The male reproductive organs—epididymis, vas deferens, seminal vesicles, and ejaculatory ducts—then develop out of the Wolffian ducts.

The external male genitalia (penis, penile urethra, and scrotum) do not emerge until around the ninth week of gestation (see Figure 7.5). Their development depends upon androgens—testosterone (produced by the Leydig cells) and its derivative dihydrotestosterone. Because the fetal pituitary is not yet secreting luteinizing hormone (LH), placental human chorionic gonadotropin (hCG) initially stimulates androgen production in the Leydig

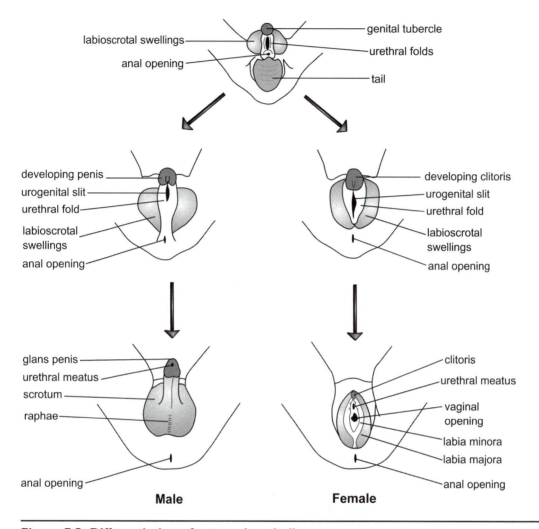

Figure 7.5. Differentiation of external genitalia.

cells. By the third trimester of pregnancy, LH production increases and exerts a greater influence over testosterone production.

During the third trimester, the testes descend from the abdominal cavity into the scrotum. This placement is important—by moving the testes outside of the body, the sperm stay cooler than the core body temperature, which keeps them viable. Also during this period, the germ cells that will evolve into mature sperm move from the embryonic yolk sac to the testes.

In the absence of testosterone and MIS, the Wolffian ducts regress and the internal female reproductive organs (uterus, fallopian tubes, and upper vagina) begin to grow from the Müllerian ducts (see Figure 7.6). Ovarian

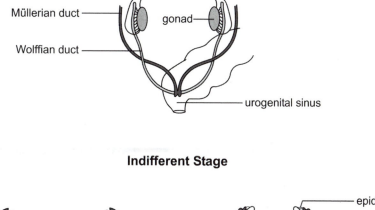

Indifferent Stage

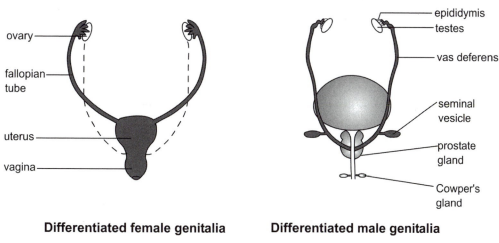

Differentiated female genitalia   **Differentiated male genitalia**

Figure 7.6. Differentiation of internal genitalia.

hormones are not necessary for the female genitalia to develop—in fact, the ovaries remain quiescent during fetal development. In females, LH and FSH levels peak near the end of gestation, then peak again at two to three months postpartum, and finally drop and remain low until puberty.

CHILDBIRTH

After the fetus has developed for about forty weeks (gestation length varies from person to person), the uterus begins to contract, signaling the onset of labor. Scientists are still unsure as to what factors trigger labor, but they believe it may have something to do with placental corticotropin-releasing hormone (CRH), which peaks in the latter weeks of pregnancy. Placental CRH is thought to stimulate the fetal pituitary, triggering a surge in adrenocorticotropic hormone (ACTH) production. ACTH, in turn, stimulates secretion of an estrogen precursor called dehydroepiandrosterone sulfate (DHEA-S), which prepares the mother for labor (see Chapter 3).

During labor, stretching of the vagina and cervix triggers a surge in fetal and maternal oxytocin levels. Oxytocin stimulates the smooth muscle cells surrounding the mother's uterus to contract, which helps push the baby out. Doctors often use a synthetic form of oxytocin (pitocin) to initiate or speed up labor. Prostaglandins also assist in the delivery by ripening the cervix (prepare it for labor), increasing uterine motility, and maintaining labor once it has begun. Synthetic prostaglandins may also be used to induce labor.

Following delivery, the mother's hormonal balance once again shifts. Placental hormone production ceases, steroid hormone levels drop, prolactin levels fall, and FSH and LH secretions are suppressed.

Lactation

During pregnancy, prolactin levels rise, preparing the breasts for lactation. Prolactin levels are ten to twenty times their normal concentration by full gestation. Prolactin stimulates the development of milk-producing cells called alveoli and the ducts that carry milk from the alveoli to the nipples. Prolactin also initiates milk production after the child is born. Estrogen and progesterone also influence breast development. Growth hormone, insulin, and glucocorticoids influence the metabolism of glucose, amino acids, fatty acids, and calcium, which are necessary for milk production.

Although the breasts are preparing for lactation, milk release is suppressed during pregnancy by high circulating levels of estrogen and progesterone. A day or two after delivery, levels of these hormones drop, allowing milk production to begin. When the infant suckles, sensory neurons in the mother's nipple send impulses to the hypothalamus, triggering oxytocin secretion. Oxytocin causes the smooth muscles surrounding the alveoli to contract, pushing the milk into the ducts (see Figure 7.7). The ducts carry milk to the nipples (this is called let-down). Suckling also stimulates prolactin release, which maintains milk production.

PUBERTY

The anterior pituitary begins secreting the hormones that orchestrate puberty, luteinizing hormone (LH) and follicle-stimulating hormone (FSH), around the sixth or seventh week of gestation. Levels of these hormones peak once between the twenty-fifth and twenty-ninth weeks of pregnancy, then peak again between the baby's third month and second year. After age 2, LH and FSH levels drop again and remain suppressed until puberty, inhibited by small amounts of circulating sex steroids.

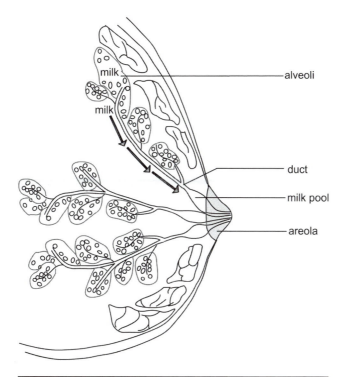

milk

milk

milk

alveoli

duct

milk pool

areola

Figure 7.7. Lactation.
Milk is produced in the alveoli and carried to the nipple via a system of ducts through a process called let-down.

After several years of relative quiet, the hypothalamus begins working harder. It releases stronger and more frequent pulses of gonadotropin-releasing hormone (GnRH), which stimulates the anterior pituitary to release LH and FSH. In addition, gonadotropin secretion becomes less sensitive to negative feedback inhibition by circulating sex steroids, as it was in early childhood.

Even before the onset of puberty, boys' bodies begin to produce androgens. By about age 7 or 8, the male adrenal gland starts to pump out dehydroepiandrosterone (DHEA) and androstenedione. Adrenal androgens trigger the part of puberty called **adrenarche,** which is characterized by a growth spurt as well as pubic and body hair growth. Closer to puberty, which usually occurs between ages 13 and 15 in boys, secretion of LH and FSH increases. LH stimulates the Leydig cells of the testes to secrete testosterone, and FSH stimulates the Sertoli cells to produce sperm (**spermatogenesis**).

During the next couple of years, a number of androgen-driven changes occur that transform the boy into a man (see Figure 7.8):

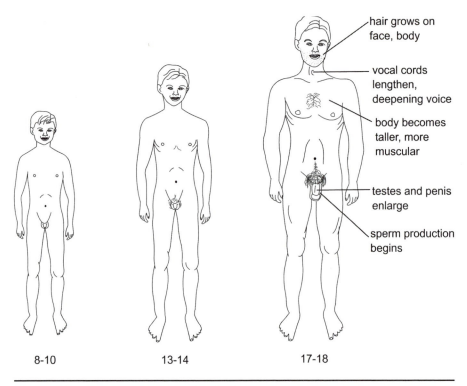

hair grows on
face, body

vocal cords
lengthen,
deepening voice

body becomes
taller, more
muscular

testes and penis
enlarge

sperm production
begins

8-10   13-14   17-18

Figure 7.8. Changes in the male during puberty.

- The testes and penis grow, and the testes become darker and thicker.

- Hair grows on the face, body, and pubic area.

- The larynx enlarges and the vocal cords become longer and thicker, deepening the voice.

- The sweat glands mature, causing a body odor.

- Acne may appear on the face and body.

- Height and muscle mass increase (because of testosterone, boys wind up with nearly twice the muscle mass of girls).

In females, puberty usually occurs between the ages of 9 and 16. As is the case in males, the female adrenal gland begins to secrete weak androgens years before puberty actually begins. These androgens, as well as androgens from the ovaries, cause hair to grow on the body and pubic area (adrenarche). Just before the onset of puberty, GnRH secretion increases, triggering a rise in LH and FSH secretion. These hormones stimulate the development of ovarian follicles and initiate estrogen and progesterone production (see Chapter 6). As more of these sex hormones are produced, the female body begins to change. The hips become curvier as fat redistributes,

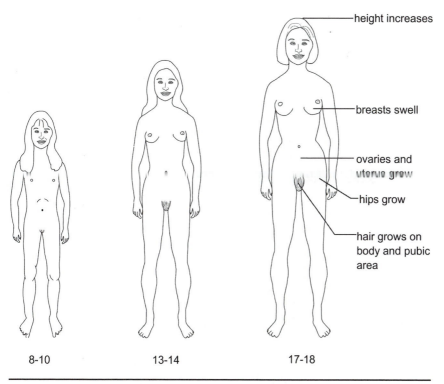

height increases

breasts swell

ovaries and
uterus grow

hips grow

hair grows on
body and pubic
area

8-10   13-14   17-18

Figure 7.9. Changes in the female during puberty.

the breasts become rounder and fuller (called **thelarche**), the height increases, and the reproductive organs enlarge (see Figure 7.9).

About two years after the breasts begin to grow (usually between the ages of 11 and 14), girls experience their first menstruation (**menarche**), or shedding of the uterine lining. Ovulation—the passage of a ripened egg from the ovarian follicle through the fallopian tube—does not typically occur until several months later. Once estrogen feedback mechanisms have matured, they stimulate the LH surge that initiates ovulation (see Chapter 6).

Both sexes experience a growth spurt during puberty, driven by increased secretion of growth hormones and sex steroids. Girls usually begin their growth spurt earlier than boys, which is why many junior high school pictures show girls towering over their male classmates. Boys lag about two years behind girls, but ultimately they gain more height (boys can grow nearly a foot during their growth spurt). Sex steroids influence growth indirectly—by increasing the secretion of growth hormone and stimulating the production of insulin-like growth factor 1 (IGF-1) in cartilage—and directly by influencing the maturation of bone-building cells called osteoblasts, and cartilage cells called chondrocytes. Finally, sex hormones

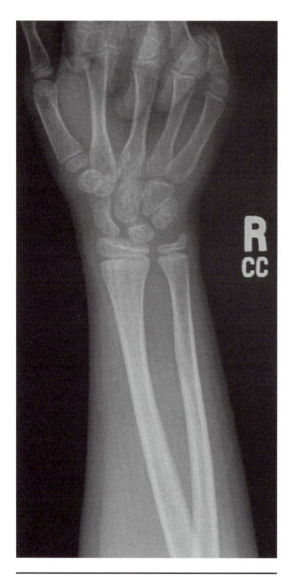

An x-ray showing growth plates on a child's arm. © Custom Medical Stock Photo.

promote closure of the epiphyses (plates) in the long bones, which ultimately stops bone growth (see photo). Because girls tend to begin their growth spurt earlier than boys, their epiphyses close earlier, which is why most high school pictures show teen boys towering over their female classmates. In the event that either growth hormones or sex steroids are deficient, growth will usually be stunted (see Chapter 10). After puberty, levels of growth hormones and sex steroids drop and remain at normal adult levels throughout adulthood.

AGING AND THE ENDOCRINE SYSTEM

With advancing age, the body loses many of its hormone receptors, and hormone production shifts to maintain homeostasis while other body functions are changing. But despite the alteration in hormone balance, hormones do not directly influence the aging process.

Menopause

Between the ages of 40 and 60, women go through a hormonal change of life that results in the end of menstruation, the cessation of estrogen production, and the loss of reproductive ability. Menopause ends with the final menstrual period, but it can begin several years earlier. Most women have their last period at around age 50, but women as young as 40 may begin to show symptoms of menopause.

During her childbearing years, a woman will have a menstrual cycle about once every twenty-eight days, during which an egg is released from the ovarian follicle and sent down the fallopian tube for possible fertilization (see Chapter 6). Women are born with a limited number of ovarian follicles (about a million), which are eventually depleted. By menopause, a woman may only have a few thousand or a few hundred follicles remaining. Most of these follicles are lost not during ovulation, but through cell death.

The first outward sign of menopause is an irregular menstrual cycle. Menstrual periods, which usually arrive at the end of the twenty-eight-day cycle, become fewer and farther apart during the initial period of menopause (called perimenopause), and women may start to experience hot flashes and other symptoms. The menstrual cycle becomes more irregular because the few ovarian follicles that remain do not respond well to FSH and LH stimulation, and thus produce less and less estrogen and progesterone. The amount of menstrual flow also decreases, and the duration of each period becomes shorter. Fertility fluctuates during this period, but conception can still occur until a woman has her final period. Eventually, the ovaries stop producing enough estrogen to build the uterine lining. Egg release and menstruation stop, resulting in the end of fertility. Menopause is not complete until menstruation has been absent for at least twelve months.

Without estrogen to provide negative-feedback inhibition, levels of FSH and LH in the blood surge. Following menopause, FSH and LH levels steadily decline. The adrenal gland continues to secrete small amounts of estrogen and progesterone, and it becomes the primary source of these hormones.

During menopause, a drop in estrogen levels can cause the following symptoms:

Vaginal changes: Estrogen loss can leave the vagina drier and thinner, making intercourse uncomfortable or even painful.

Hot flashes: Skin temperature rises, causing a rush of heat to the face, neck, and upper body that can last anywhere from thirty seconds to five minutes. Some women are awakened by hot flashes while they sleep. These night sweats, as they are called, are characterized by a flush of intense heat followed by chills.

Cardiovascular disease: Without estrogen, cholesterol and LDL levels rise to unhealthy levels, increasing a woman's risk of developing heart disease.

Masculine characteristics: Estrogen levels drop following menopause, even as the female body continues to produce small quantities of androgens. These male hormones sometimes cause hair growth on the face, chest, and abdomen.

Thinning of the bones: Without estrogen, the rate of bone loss increases and the bones become weak. Bone loss over time can lead to osteoporosis.

Emotional swings: During menopause, women may feel more anxious or irritable due to hormonal changes and age-related life changes (for example, the death of a loved one or empty nest syndrome).

Hormone Replacement Therapy (HRT)

Menopause can prove a difficult time for women—both emotionally and physically. During this change of life, levels of the hormones estrogen and progesterone drop dramatically (see the discussion on Menopause earlier in this chapter). With declining hormone levels come a number of uncomfortable symptoms, including hot flashes, vaginal dryness and irritation, difficulty sleeping, mood swings, and increased urinary infections.

Menopause has also been associated with less overt physical changes. The risk of heart disease, the leading killer of women in the United States, rises steeply following menopause. A woman's rate of bone loss (**osteoporosis**) also increases during this period. When the amount of bone lost exceeds that which is replaced (see the discussion on bone remodeling in Chapter 4), the bones become more susceptible to fractures and breaks.

In the mid-twentieth century, scientists began to realize that by replacing naturally produced estrogen and progesterone (see Chapter 6 for a discussion of these hormones) with hormone therapy, they could alleviate the discomforts and protect against the health risks associated with menopausal hormone loss. The first **hormone replacement therapy,** Premarin, made its debut in 1942. Many doctors touted hormone therapy as a veritable "fountain of youth" that would slow the aging process and make women feel renewed and revived. In the decades to follow, a number of studies emerged that noted several benefits to hormone therapy. According to the studies, hormone therapy relieved the symptoms of menopause, prevented bone loss and fractures, and protected against heart disease. By the mid-1970s, estrogen was one of the top five drugs prescribed in the United States. Two decades later, about 6 million American women were using combination hormone therapy at any given time.

But beginning in the 1970s, a new wave of studies began raising questions about the effectiveness—and safety—of hormone replacement therapy. In 2002, a landmark study on combination hormone therapy (estrogen plus progestin) called the Women's Health Initiative (WHI)—a multicenter, eight-year clinical study involving thousands of U.S. women, which examined the role of diet, exercise, calcium supplements, and hormone replacement therapy in regard to heart health—was halted because of safety concerns. The study's findings have only fueled the growing controversy over hormone replacement therapy. (See the "Risks Associated with Hormone Replacement Therapy" section later in this chapter.)

What Is Hormone Replacement Therapy?

Hormone replacement therapy (HRT) replaces the natural hormones a woman loses during menopause. Estrogen can either be taken alone (called estrogen replacement therapy or ERT) or in combination with the synthetic form of progesterone (called estrogen-progestin therapy [EPT] or hormone replacement therapy [HRT]). With estrogen therapy, the circulating level of estrogen in a woman's system roughly doubles, although it still does not reach the premenopausal level. Because long-term use of estrogen has been linked to an increased risk of endometrial cancer (cancer of the uterine lining), women who have not had a **hysterectomy** are typically given combined estrogen-progestin therapy. Progesterone is thought to prevent the unnatural growth of cells lining the uterus.

The type of estrogen most commonly used in therapy is conjugated equine (made from the urine of pregnant horses), but newer forms are made from plants. "Natural" progestins are also derived from plants; others are synthetically produced.

Hormones may be administered in one of several ways: pills, injections, skin patches, or vaginal creams or rings (see Table 7.2). Vaginally administered hormones tend to be in lower concentration than those given orally. The form in which hormones are taken depends upon the type of treatment needed. For example, vaginal estrogen creams can ease dryness by acting

TABLE 7.2. Types of Medications Used in Hormone Therapy

Hormone Type	Medication Names	What It Does
Estrogen	Premarin—Orally administered estrogen hormones (also can be administered in a low-dose vaginal cream)	• Reduces hot flashes. • Treats vaginal dryness, itching, and burning. • Reduces osteoporosis risk.
Progestin	Depo-Provera—Injection that contains a progesterone called medroxyprogesterone acetate	• Prevents the overgrowth of cells in the uterine lining (in women who have not had a hysterectomy).
Estrogen plus progestin	Prempro—Oral combination of estrogen and progestin	• Reduces hot flashes. • Treats vaginal itching, dryness, and burning. • Reduces osteoporosis risk.
Other estrogen delivery systems	Estring—Estrogen vaginal ring	• Soothes vaginal dryness and irritation.
	Alora, Climara—Estrogen skin patch	• Relieves hot flashes and vaginal dryness.
	Estrace—Vaginal cream	• Relieves vaginal dryness and irritation.
Other progestin delivery systems	Crinone—Natural progesterone vaginal gel	• Treats vaginal dryness and irritation.

upon the vaginal epithelium (the thin layer of tissue that covers the inside of the vagina), whereas pills more often target other menopause symptoms, such as hot flashes. Hormones may be taken daily (continuous use) or on certain days of the month (cyclic use). In the past, most women remained on hormone therapy for a period of two years or more, but in light of the WHI, the U.S. Food and Drug Administration (FDA) has begun recommending that estrogens and progestins be given for the shortest period of time, and at the lowest dose possible, to achieve the desired results.

Benefits of Hormone Replacement Therapy

Hormone therapy has been used to combat the symptoms of menopause and restore patients' quality of life. The treatment has been recommended for women who have experienced severe discomfort with menopause or who are at a high risk for osteoporosis.

Some of the benefits associated with hormone therapy are:

> *Menopausal Symptoms*: Hormone therapy can relieve hot flashes, sleep disturbances, as well as vaginal itching and dryness associated with post-menopausal atrophy. Hormones may also help women who suffer from emotional distress related to menopause.

> *Osteoporosis*: As mentioned above, the rate of bone loss rises sharply during menopause. Therapy with estrogen, or estrogen plus progesterone, can slow bone loss and protect against fractures of the hip, vertebrae, and other bones.

> *Heart Disease*: Early evidence indicated that estrogen reduced the risk of coronary heart disease in postmenopausal women because it raised the level of "good" cholesterol (high-density lipoproteins, or HDLs) while it decreased the levels of "bad" cholesterol (low-density lipoproteins, or LDLs).

Risks Associated with Hormone Replacement Therapy

Today, the benefits of HRT are a subject of intense debate. Despite early studies that suggested HRT reduced a woman's risk of heart disease, emerging evidence has indicated the exact opposite. As part of the WHI trial, researchers compared the effect of HRT taken as combination therapy (0.625 milligrams of estrogen and 2.5 milligrams of progestin) with a placebo in 16,600 healthy women with a uterus who were between the ages of 50 and 79. (Prempro is the most commonly prescribed version of this popular combination. However, the combination may also be taken as two separate pills: an estrogen such as Premarin and a progestin like Provera. The study did not evaluate HRT taken in other doses or forms.) The theory, based on earlier evidence, was that taking this combination of HRT would result in a decrease in coronary heart disease.

In July 2002, a little over five years into the study, the arm of the trial involving HRT was halted when overwhelming evidence revealed that the risks of HRT (especially of cardiovascular disease and invasive breast cancer) outweighed the benefits of lowering the risk of colon cancer and bone fractures. According to the American College of Obstetricians and Gyne-cologists (ACOG), the study found that if 10,000 women took combination HRT for one year:

- Seven will have a heart attack or other coronary problem.
- Eight will suffer a stroke.
- Eighteen will develop serious blood clots.
- Eight will develop invasive breast cancer.

After five years of HRT, women had a 26 percent increase in invasive breast cancer. While the risk for breast cancer also jumped in the placebo group (advancing age increases one's risk for breast cancer), the risk for the HRT group appeared to increase at a higher rate than normal. ACOG says that while the rate of increased breast cancer risk may seem small, in a drug taken by millions of women over many years, the risk can result in a large number of new breast cancer cases. Further follow-up studies indicate an even larger risk of breast cancer among HRT patients. In August 2003, The *Lancet* reported on a British study that suggested HRT doubled a woman's risk of breast cancer.

According to the NIH, the study also revealed that the risk of blood clots in the first two years of therapy is four times higher than in those who do not take HRT. In addition, an HRT patient's stroke risk gradually increases after two or more years. The full study report was published in the July 17, 2002, issue of the *Journal of the American Medical Association* (*JAMA*).

Once believed to be cardio-protective, the WHI findings prove that the es-trogen plus progestin combination studied in WHI does not prevent heart disease. The study concluded that women should not take (or continue to take) combination HRT for the prevention of heart disease. The FDA has never approved estrogen and progestin hormones for warding off heart dis-ease, although doctors are allowed to prescribe them for "off-label" use. An-other study published in the August 2003 edition of the *New England Journal of Medicine* added to the mounting evidence that combination HRT boosted a woman's risk of a heart attack.

The pool of literature regarding the bad effects of HRT continues to roll in. Results from another branch of the WHI study showed that combination HRT also has a significant impact on a woman's mental abilities. In post-menopausal women age 65 and older, combination HRT doubled a woman's risk for developing dementia, including Alzheimer's disease. Other studies continue to show that HRT increases a women's risk for gallbladder disease.

Figure 7.10 illustrates the differences in risks and benefits revealed dur-ing the HRT study. The WHI findings apply broadly; researchers noted no differences in risk by prior health status, age, or ethnicity. The study results do not apply to postmenopausal use of estrogen alone. That arm of the study,

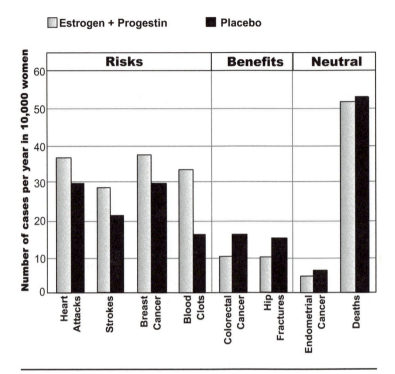

Figure 7.10. Disease rates for women as shown by the WHI study.

called the ERT study, is continuing because the women taking estrogen alone have not experienced the same elevated risks of cancer or cardiovascular disease. Patients in the ERT study take a daily 0.625 mg dose of conjugated equine estrogen (Premarin).

In response to the WHI findings, the U.S. FDA has ordered labeling revisions for Prempro, Premphase, and Premarin for patients and physicians. The new warning, the highest level of warning information in labeling, emphasizes the increased risks for heart disease, heart attacks, strokes, and breast cancer. This warning also strongly points out that these products are not approved for the prevention of heart disease. The FDA has also adapted the approved indications to clarify that these drugs should only be used when the benefits clearly outweigh risks. The approved use for the treatment of moderate to severe hot flashes associated with menopause has not changed. The two revised indications are as follows:

> *Indication*: Treatment of moderate to severe symptoms of vulvar and vaginal dryness and irritation associated with the menopause. *Change*: Topical products should be considered when HRT is prescribed solely for this purpose.

Indication: Prevention of postmenopausal osteoporosis. *Change*: When prescribed solely for the prevention of weak bones, estrogens and combined estrogen-progestin products should only be considered when the significant risk of osteoporosis outweighs the risks of the drug.

Why the change of heart concerning HRT? Previous observational studies compared women who took HRT with those who did not. Those studies, however, were not as sophisticated as the double-blinded WHI study, which compared HRT patients with a placebo control group with neither group aware of which pill they were taking. There is no doubt it will take a long while for researchers to fully untangle the multitude of study results. Obviously, the decision for a woman to use, or continue to use, HRT has become considerably complex. Experts advise women to talk to their doctor and consider carefully the advantages and disadvantages of discontinuing hormone therapy. In some cases, halting therapy may do more harm than good. For example, a study published in the November 2002 issue of *The Journal of Clinical Endocrinology & Metabolism* (*JCEM*) found a rapid decrease in bone mineral density in elderly women who halted treatment. The most bone loss occurred during the first year following treatment.

The best way to halt therapy is still being evaluated. When stopping HRT cold turkey, some patients may experience heavy vaginal bleeding. If menopausal symptoms recur with this abrupt approach, a gradual approach may be advised.

Alternatives to Hormone Replacement Therapy

Many dietary supplements, natural medications, and lifestyle changes may relieve the symptoms of menopause without the risks of hormone replacement therapy:

- Soy, as well as several types of herbs (including black cohosh, wild yam, and dong quai), contain plant-derived estrogens called phytoestrogens. These naturally occurring estrogens may offer relief from hot flashes, but potential side effects are still unknown. (However, a 2003 Finnish study published in the journal *Obstetrics & Gynecology* found that pure phytoestrogens did not alleviate menopausal symptoms in breast cancer patients.) Experts caution that so-called alternative therapies are not regulated by the FDA and do not go through the same stringent approval process as traditional medications.

- Antidepressants such as fluoxetine (Prozac) and paroxetine (Paxil) have been found to reduce the frequency of hot flashes.

- A new class of medications, called selective estrogen receptor modulators (SERMs), offers some of the benefits of estrogen therapy but doesn't carry an increased risk of breast and uterine cancer. Nolvadex (Tamoxifen), the first approved SERM, is used to prevent breast cancer in high-risk women (but it may increase the risk of uterine cancer); raloxifene (Evista) decreases the risk of bone fractures and increases bone density.

- Calcium plus vitamin D can strengthen bones, thereby lowering the incidence of fractures related to osteoporosis, while biphosphonates such as risedronate (Actonel) and alendronate (Fosamax) fight bone loss.

- Exercise, diet, and a healthy lifestyle (i.e., avoiding smoking and excessive consumption of alcoholic beverages) can prevent heart disease and regulate mood during menopause.

Aging and the Male

Although men do not officially go through menopause, they do experience subtle changes associated with gradual alterations in hormone production. At around age 40, testosterone levels begin to decrease as the testes become less sensitive to LH stimulation (see Chapter 6). The decline in male hormone production is sometimes referred to as "andropause." As testosterone levels drop, the levels of a protein called sex hormone-binding globulin (SHBG) rise. SHBG, which carries testosterone through the bloodstream, traps testosterone and prohibits it from reaching its target tissues. The testosterone left over is referred to as "bioavailable." This limited supply of testosterone is free to act upon tissues. As testosterone declines, it exerts less negative-feedback inhibition on blood LH and FSH levels, which subsequently rise. Sperm production also declines slightly, although most men can still father children late in life.

Symptoms associated with reduced testosterone in older males include:

- Loss of muscle mass and tone, and increase in body fat
- Decreased libido
- Irritability, depression, and mood swings
- Fatigue
- Erectile dysfunction
- Hot flashes
- Loss of mental acuity

8

History of Discovery

Unlike some other systems of the body, thorough knowledge of the endocrine system's existence and function evolved at a snail's pace. Although observations of endocrine malfunction date back to ancient times, the "birth" of clinical endocrinology did not occur until many centuries later, in the 1800s. Since that time, discoveries and advancements have progressed at an exceptionally rapid rate. This chapter examines the developments in endocrinology from the beginning of time until the twenty-first century.

EARLIEST HISTORY

In ancient times, the common practice of castration caused endocrine malfunction. Medical observers noted changes in sexual characteristics and fertility in men and animals that had been castrated. The underlying mechanisms, however, would not be understood for many centuries to come.

One of the first endocrine disorders to be documented comes from ancient China (1600 BCE), where doctors were believed to have treated goiter (see Chapter 11) with a concoction of burnt sponge and seaweed. The incidence of goiter among the Chinese was widespread from the fifth century forward. In fact, the great explorer Marco Polo makes reference to the disease in Chapter 32 of his celebrated book, *The Travels of Marco Polo*. In it he writes, "Of the Province of Karkan, the Inhabitants of which are Troubled with Swollen Legs and with Goitres." The Greeks believed that goiter resulted from drinking snow-water; the earliest Romans held fast to a similar theory that the enlarged thyroid was caused by water impurities. Roman military commander Pliny the Elder, also known as Gaius Plinius Secundus (23–79 CE), noted in his writings, "Only men and swine are subject to swellings in the throat, which are mostly caused by the noxious quality of the water they drink." The Hindus described goiter as being characterized by an internal prickling pain.

Cases of stunted and rapid growth were also evident in early societies. In ancient China, dwarfs were regarded as jesters. In Egypt, the god Bes is frequently portrayed as a dwarf. Pliny's writings refer to races of dwarfs in Asia and Africa. He was correct; today African pygmies, for example, rise only to a height of about four and a half feet. In the Bible (Leviticus, 21:17–20), dwarfs were considered misfits and were denied access to the temple. Yet during the Greco-Roman period (156 BCE–576 CE), the Roman Empire was very keen on having cretinous (see Chapter 11) dwarfs, perhaps to serve as royal jesters similar to the Chinese.

The Greeks and Jews both make reference to gigantism. Goliath of Gath is a famous biblical giant. According to I Samuel 17:4–7, Goliath stood at "six cubits and a span . . . and the weight of his coat was five thousand shekels of brass." Giants are also mentioned in the story of Noah in Genesis 6:4: "There were giants in the earth in those days." Tribes of giants are noted throughout the Old Testament: Numbers 13:32–33 notes, "All people that we saw . . . are men of great stature." In the second book of Samuel, battles resulted in the death of four Philistine giants; II Samuel 21:15–22 describes a giant man with twenty-four fingers and toes. In his magnum opus, *An Introduction to the History of Medicine* (1929), F. H. Garrison (1870–1935) suggests that biblical giant was actually an acromegalic. Many experts, however, have disputed this claim.

Acromegaly appears prevalent in the early Roman culture as well. In their book, *The Pharaoh Akhenaten: A Problem in Egyptology and Pathology* (Bulletin, of the History of Medicine 26, 293, 1963), authors C. Aldred and A. T. Sandison suggested that Akhenaten might have suffered from a pituitary tumor, which resulted in the pharaoh's acromegalic face and obesity.

Cases of acromegaly have been traced back to the first Norse settlers in Greenland. In his publication *New Discoveries Relating to the Antiquity of Man* (1931), Sir Arthur Keith (1866–1955) diagnoses a twelfth-century skull found at the southern tip of Greenland as acromegalic. Several Nordic tales make reference to races of unusual stature and appearance.

The most important physician in ancient Greco-Roman times was Galen (130–200 CE). The Greek physician had a profound influence on medical theory and practice until the mid-seventeenth century and is often touted as the greatest medical thinker of ancient times after Hippocrates. He believed that the pituitary drained phlegm from the brain into the nasal cavities, a theory that remained intact for 1,500 years, until Conrad Victor Schneider (1614–1680) refuted it in 1660. Galen also theorized that the thyroid gland filtered blood and that its secretions lubricated the voice box.

Diabetes mellitus (see Chapter 12) appears to have been frequently studied by many ancient and medieval physicians, although the disease's myriad of symptoms stirred confusion among them. Galen believed the disease to be a weakness of the kidneys. The "Chinese Hippocrates," Chang Chung-ching (c. 160–c. 219 CE), was one of the first to study diabetes. The first Chinese physician to make note of diabetes' trademark sign, "sweet" urine, was Chen Chhuan, who died in 643 CE. (The Hindus also made reference to sweet urine.) In one of his published works, Chhuan noted three forms of diabetes. He described the first form as being marked by extreme thirst and by frequent, large volumes of urine. In the second form, Chhuan's patients were overweight, voracious eaters, but had little thirst and produced lower volumes of urine (although they urinated frequently). In the last form, he noted a thirsty patient who had swelling of the extremities, necrosis of the feet, impotence, and frequent urination. Twentieth-century London physician Joseph Needham (1900–1995) identified Chhuan's first form as diabetes mellitus.

The first complete clinical description of diabetes mellitus in Europe is credited to the Greek Aretaeos of Cappadocia (second century CE). He called the disease a "wonderful affection, not very frequent among men, being a melting down of the flesh and limbs into urine." Aretaeos goes on to note the extreme thirst typical of diabetes today: "Life is short, disgusting and painful, thirst unquenchable, death inevitable."

THE SIXTEENTH-CENTURY RENAISSANCE

Aureolus Theophrastus Bombastus von Hohenheim (1493–1541), better known as Paracelsus, is often regarded as the most important medical observer of the sixteenth century. Like many of his predecessors, he believed that bad drinking water led to goiter. He blamed iron deposits for the outbreak of the disease. The Swiss physician and former military surgeon was the first to observe the connection between cretinism, endemic goiter, and congenital **idiocy** (severe mental retardation). He conducted firsthand experiments on endemic goiter in the Duchy of Salzburg. Paracelsus's notes on cretinism are found in his *De Generatione Stultorum* (1567). His unbiased study of diabetes mellitus revealed it to be a severe generalized disease, a conclusion that opposed Galen's findings. (Galen had theorized that diabetes mellitus was a weakness of the kidneys.) Paracelsus's urine evaporation experiments resulted in the formation of considerable amounts of a solid. Following his research,

Woodcut portrait of Paracelsus. © National Library of Medicine.

Paracelsus called for closer examination of diabetic urine.

Paracelsus's observations on cretinism and goiter secured him a place in medical history. Following Paracelsus's important discoveries, numerous physicians launched goiter studies in endemic communities such as the Valais, Styria, and the Pyrenees. Dutch doctor Pieter van Foreest (?–1597) noted an abundance of cretins on the Italian side of the Swiss border.

In the 1500s, interest in the thyroid gland grew. Andreas Vesalius (1514–1564) described the thyroid as two glands poised on each side of the larynx. A few years later, Giulio Casserio (1561?–1616) similarly proposed that the thyroid was a single organ split in two. Bartolomeo Eustachi (Eustachius: 1520–1574), best known for his discovery of the Eustachian tube in the ear, referred to the thyroid gland as "glandula laryngea."

THE SEVENTEENTH-CENTURY MICROSCOPISTS

By the turn of the seventeenth century, scientists had identified the thyroid and the thymus, salivary, and lymphatic glands. Yet gland structure went unobserved until the advent of Galileo Galilei's (1564–1647) microscope in 1609. Faced with the realization that early microscopes failed to give clear pictures, mechanical genius Robert Hooke (1635–1703) created one of the first compound microscopes. In 1665, he published the results of his observations with the new scope in his *Micrographia, or Some Physiological Descriptions of Minute Bodies Made by Magnifying Glasses; with Observations and Inquiries Thereupon.*

Italian professor Marcello Malphigi (1628–1694) is regarded as the greatest figure among the microscopists. He made fundamental microscopic discoveries about the lungs as well as the capillaries within them. In Messina, where he served as a professor of medicine, Malphigi studied gland structure, referring to a gland as a "secretion machine." He also coined the phrase "corpus luteum," referring to the glandular tissue of the ovary that forms at the site of a follicle rupture after ovulation.

Malphigi's interested in glandular studies stemmed from J. George Wirsung's (1589–1643) discovery of the human pancreatic duct in 1631. Today this main duct, which collects pancreatic juices and empties into the duodenum, is referred to as the duct of Wirsung. Despite his namesake, Wirsung never had a chance to publish his remarkable discovery because a Belgian student murdered him in 1643. Anatomy professor Johann Vesling (1598–1649) later announced the discovery.

German physician Franciscus Sylvius (1614–1762) was the first to suspect that the pancreas discharged a "juice" into the intestines; an observation that, indeed, was correct. The founder of the seventeenth-century Iatrochemical School of Medicine, Sylvius helped transform medical emphasis from mystical speculation to rational, concrete mechanisms based on the universal laws of physics and chemistry. He was the first to distinguish between **conglomerate glands** (glands that collect together into a ball or mass) and **conglobate glands** (glands that form into a round compact mass).

Twenty-two years after Wirsung's duct discovery, Dutch physician Regnier de Graaf (1641–1673) studied the function of the endocrine pancreas. In his work *Disputatio medica de natura et usu succi pancreatici* (1664) de Graaf describes his six attempts to collect pancreatic juice from dogs. His experiments, however remarkable, failed to produce significant results, due to the lack of chemistry knowledge at the time. (Claude Bernard [1813–1878] would launch similar studies in the nineteenth century.) In his short thirty-two years, de Graaf performed an amazing number of medical experiments. His observations on the gonads and pancreas landed him a position among the greatest scientists of all time. De Graaf documented the structure of the testis and performed vasectomies. His most significant observation was that the testis was not glandular; instead, it consisted largely of tubules, which he called vascular seminalia. De Graaf was the first to highlight the ovarian follicle from which the mature egg is released, which now bears his name.

Englishmen Thomas Willis (1621–1675) is another landmark figure on the seventeenth-century medical canvas. While his colleagues focused on pancreatic function and glandular structure, Willis experimented with the urine of diabetes mellitus patients. Willis discovered diabetic urine had a sweet taste. In his *Pharmaceutice rationalis sive diatriba de medicamentorum operationibus in human corpore* (1675), he claimed to have differentiated between diabetes mellitus and diabetes insipidus, although scholars say that he simply paved the way for others to do so. Willis is believed to have been the first to treat diabetes mellitus with undernourishment and lime water.

Oddly enough, the idea of "sweet urine" gave comedy writers plenty of fodder. In his *Le Médecin malgré lui,* the great French comedy writer Molière (1622–1673), has his valet, Sganarelle, pretend to be a doctor, taste urine, and note its sweetness.

In 1672, Johann Conrad Brunner (1653–1727) nearly stumbled onto the cause of diabetes mellitus when he surgically removed the spleen and part of the pancreas from a dog. He kept the test subject alive and monitored the animal for symptoms. Brunner noted extreme thirst and large volumes of urine (polyuria). The connection between the disease and the pancreas, however, went undocumented. Scholars theorize that Brunner's dog survived because some of its pancreas remained intact.

For centuries, physicians and medical observers used the word "gland" indiscriminately. But in 1656, Cambridge physician Thomas Wharton (1614–1673) offered the first accurate definition. In his *Adenographia: sive glandularum totius corporis descriptio* (1659), Wharton categorized the following as glands: lymphatic, pancreas, thymus, salivary, thyroid, pineal, and pituitary. He discovered the duct of the submaxillary salivary gland, which is today referred to as Wharton's duct. Wharton described the thyroid more accurately than any of his predecessors, but derived its modern name from an incorrect observation (Medvei, 1993). He named the gland "thyreoidea," Greek for oblong shield because its shape resembled the shields used in Ancient Greece. However, it is actually butterfly-shaped. Wharton went on to describe three functions of the thyroid: (1) to draw moisture from the recurrent nerve, (2) to "cherish" the cartilages, and (3) to oil the larynx.

THE EIGHTEENTH CENTURY

The eighteenth century produced a number of important medical observers, all of whom had a great impact on endocrinology. Topping the list of important physicians are Théophile de Bordeu (1722–1776) and Albrecht von Haller (1708–1777).

Bordeu theorized that each organ of the body gave off hormones, an idea very close to present-day thinking. In his 1752 publication, *Recherches Anatomiques sur le Différentes Positions des Glandes et sur leur Action*, he presented a modern description of the glands that later became known as endocrine organs. In 1753, his thesis on tuberculous glands netted him the prize of the Académie de Chirurgie. Two years later, in his last work on chronic diseases, he suggested that all bodily organs discharged secretions into the bloodstream. He is believed to have detected tiny tubes running from the thyroid to the trachea, suggesting that the thyroid was a true gland with excretory ducts. De Bordeu continued to make remarkable observations until the day he died; in fact, he predicted the actual date of his death: November 24, 1776. In 1903, Austrian professor, Max Neuburger (1868–1955), who discovered de Bordeu's writings, proclaimed the late physician the "Father of Endocrinology." Despite little regard for his work, de Bordeu more closely approached the modern theory of internal secretion than any other researcher and greatly influenced science in the following years.

Albrecht von Haller, the other eighteenth-century medical giant, is regarded as the greatest systematist since Galen. His encyclopedia, *Elementa Physiologiae Corporis Humani* (Physiological Elements of the Human Body; 8 vol., 1757–1766), is considered a landmark in medical history. His description of the thymus, thyroid, and spleen as ductless glands influenced nineteenth-century physicians and prompted the twentieth century adoption of the term "vascular."

Haller's research also contributed to the understanding of ovarian function. His experiments revealed that the corpus luteum formed from a thickening of the ovarian wall after a follicle ruptured. Precocious puberty due to adrenocortical hyperplasia (see Chapter 12) was another topic of interest to Haller. Although he failed to understand the underlying reasons for the problem, he urged the scientific community to consider it an authentic medical condition rather than a freak occurrence.

A remarkable number of discoveries and observations occurred during the eighteenth century. In 1786, Caleb Hillier Parry (1755–1822) made what is said to be the first classical description of exophthalmic goiter. In a posthumous publication, *Enlargement of the Thyroid Gland in Connexion with Enlargement or Palpitation of the Heart* (1825), Parry describes a patient with goiter, palpitations, and protruding eyes. He also remarks: "My attendance . . . first suggested to me the notion of some connexion between the malady of the heart and the bronchocele." Other physicians, including Giuseppe Flajani (1741–1808) and Antonio Testa (1756–1814), observed cases of thyroid enlargement with palpitations, but neither realized the connection.

The late 1700s produced several medical observers interested in endocrine function. Sir Astley Paston Cooper (1768–1841) and Thomas Wilkinson King made significant strides in understanding the thyroid. Cooper, a pioneer in vascular and experimental surgery, announced in his lectures that the thyroid gland was ductless. He also performed macroscopic examination of the testis, the results of which he published in his major treatise, *Observations of the Structure and Diseases of the Testis* (1830). King's experiments on the thyroid gland revealed that its secretions passed into the lymph glands and subsequently into large veins. He writes:

> The most important novel fact concerning the thyroid gland is doubtless this, that its absorbent vessels carry its peculiar secretion to the great veins of the body, and the most simple and satisfactory method of demonstrating this fact is to expel the contents of the healthy gland by repeated and gentle compressions.

Endemic goiter continued to be a topic of conversation in the late eighteenth century. Edward Browne (1644–1708), an English physician, noted an abundance of goiter during his travels in the Styrian Alps. "Many hear [sic] have great throats, some as big as there [sic] heads," he notes. A new theory surrounding goiter emerged. H. B. de Saussure (1740–1799) and François Emmanuel Fodéré (1764–1835) suggested that the enlarged thyroid was due to concentrated air instead of water impurities. De Saussure was convinced that drinking melted snow, consuming large amounts of alcohol, and eating coarse food were *not* the causes of regional goiter. He documented goiter cases only in regions under 3,000 feet above sea level—where, he emphasized, the air was stagnant and hot.

After quite a lull, interest in the pituitary gland resumed by the end of the eighteenth century. German anatomist and physiologist Franz Joseph Gall (1758–1828) described the gland as a large ganglion. Joseph Wenzel (1768–1808) associated pituitary problems with epileptic seizures. Johann Friedrich Meckel the Younger of Prussia, insisted that the pituitary secreted fluid to nourish the brain.

The eighteenth century gave birth to significant developments in diabetes mellitus and diabetes insipidus studies. Up until this time, experts debated how or if diabetes insipidus varied from the sugar form of diabetes. Johann Peter Frank (1745–1821) presented the first clinical definition of diabetes insipidus, forever separating it from diabetes mellitus by defining the disease as "diabetes without glycosuria."

In 1780, Frank proposed a yeast test for the detection of sugar in the urine. Francis Home (1719–1813) had proposed the same idea a few years earlier. Frank, like many before him, including British surgeon William Cheselden (1688–1752), observed that carbuncles and boils often plagued diabetics. Some of the most important observations of this time came from English physician Matthew Dobson (1713–1784). In 1774, Dobson performed landmark experiments on diabetic urine. In his published results, Dobson proved that the sweetness of diabetic urine was caused by sugar. He also discovered that the blood of diabetics was sweet.

In 1788, Thomas Cawley published a "singular case of diabetes, consisting entirely in quality of the urine; with an inquiry into the different theories of that disease." Cawley was the first to suggest that diabetes mellitus might result from injury to the pancreas, a theory that was repeated by several nineteenth-century scientists.

At the close of the eighteenth century, John Rollo (?–1809), a surgeon general in the English army, documented a successful meat diet for diabetic patients. English chemist William Prout (1785–1850) was the first to note diabetic coma; he also claimed to have been the first to recommend iodine treatment for goiter.

As knowledge of diabetes ensued, scientists also made stunning new observations regarding the thymus and reproductive endocrinology. Dutch physician and anatomist Govert Bidloo (1649–1713) believed that an enlarged thymus in the fetus helped hold open the space in the lungs that would fill with air after birth. (Cooper would later refute this theory in *The Anatomy of the Thymus Gland* [1832].) G. I. Pozzi (1697–1752) maintained the belief that the thymus had contracting muscle fibers, allowing it to serve as a pump for the lymphatic system. British anatomist and physiologist William Hewson (1739–1774) suggested that production of lymphocytes was a function of the thymus.

In 1785, Italian physiologist Lazzaro Spallanzani's (1729–1799) experiments on frogs, silkworms, and other animals proved that fertilization of ova occurred by live spermatozoa. His findings supported earlier theories that spermatozoa were parasites within semen. Spallanzani studied the sperm of a large variety of species, from fish to man. His zoological studies paved the way for the research of Louis Pasteur (1822–1895).

This section would not be complete without the mention of John Hunter (1728–1793), founder of pathological anatomy in England and an early advocate of medical investigation and experimentation. Hunter performed the first successful testis transplantation, from a cock to the belly of a hen. Since ancient times, changes in sexual characteristics were noted following castration; Hunter's experiments helped confirm these observations. The army surgeon noted that an individual's sexual characteristics depended greatly on the ovaries and testicles. Hunter's research also revealed that castration prevented the regrowth of antlers in stags. Hunter's transplants are believed to be the earliest recorded experiments in the field. In 1786, Hunter reported on a man who had prostate atrophy after castration. According to Sir Everard Home (1756–1832), Hunter had once suggested artificial insemination. Based on his recommendation, a husband, using a syringe, successfully performed the procedure on his wife.

During his studies, Hunter documented a substance secreted by both male and female pigeons. He termed the substance "pigeon's milk." But not until the early 1900s did scientists discover that this pigeon's milk was the hormone prolactin.

THE NINETEENTH CENTURY: THE BIRTH OF ENDOCRINOLOGY

At the commencement of the nineteenth century, scientists had essentially mapped out the structure of the entire human body. But many of the physiological processes underlying endocrine function remained a mystery. German physiologist and comparative anatomist Johannes Müller (1801–1858) described the thyroid as a blood gland but failed to reveal its true function. Müller was an important figure in the nineteenth century; he is credited with replacing medicine's philosophical methods with experiments, observation, and analysis. He published numerous studies on the development and structure of the glands and embryonic development. The German professor described his important discoveries in a textbook, *Handbuch der Physiologie des Menschen* (Elements of Human Physiology), published from 1834 to 1840. In a later version of his famous textbook, Müller noted that "true blood-glands do not differ from other organs as regards blood and lymph vessels. To this group belong the spleen, the thyroid, the suprarenals, and the placenta (Medvei 1993, 123)."

One of the most outstanding events of the nineteenth century was the discovery of iodine, an essential chemical in the human body that is concentrated mostly in the thyroid gland. In 1811, Bernard Courtois (1777–1838), a French manufacturer of saltpeter from seaweed, stumbled upon iodine in seaweed ashes. His discovery evolved after he added too much sulfuric acid to seaweed ash while manufacturing saltpeter. The result was a violet vapor that condensed to form dark crystals. Numerous experiments and investigations ensued; in 1813 the army pharmacist announced his discovery. The element was later named iodine.

Seaweed concoctions had been used to treat goiter since ancient times. Courtois's discovery suggested that perhaps it was the iodine in the seaweed that served as a therapeutic. Scientists began administering iodine to goiter patients; the Swiss physician Jean François Coindet (1774–1845) is believed to have been the first to do so. Coindet prescribed ten drops of his tincture of iodine mixed with water to adult patients, to be taken three times a day. English chemist William Prout (1785–1850), however, proclaimed he was the first to recommend iodine to goiter patients. Prout allegedly prescribed it after taking it himself to see if it was safe. In 1821, after reading about Prout's findings, the English physician Alexander Manson (1774–1840) administered iodine to goiter patients with great success.

In the meantime, Gaspard Adolphe Chatin (1813–1901) announced that iodine extracted from freshwater plants could prevent endemic goiter and cretinism. Chatin theorized that traces of iodine were universally present in nature, in varying amounts. One of his most important observations was that the iodine content of food and water was extremely low in goiter districts. He calculated that goiter-free communities in Paris consumed 1/100–1/200 mg of iodine daily; in the highly goitrous region of the Alpine valleys, the maximum intake was 1/2000 mg. He recommended a higher reliance on mineral water springs, which contained higher levels of iodine than ordinary drinking water. In 1833, J. B. Boussingault (1802–1887) suggested the iodization of common table salt to help prevent goiter. Jean-Louis Prevost (1790–1850) and Chatin supported his recommendation. However, after a French Academy of Science commission proclaimed that small amounts of iodine would not produce any physiological effect, Chatin's work was discredited. Scientists would rediscover his important findings in the twentieth century.

With the increasing use of iodine came observations of toxicity. Frederic Rilliet (1814–1861) noted in his paper *Constitutional Iodism* that even small doses could be dangerous to some patients. His description of iodine overdose was presented to the Academy of Medicine in Paris in 1850. Coindet also made mention of iodine toxicity in 1821. Years later, Hermann Lebert (1813–1878) encouraged a full clinical observation of the thyroid; investigations did not begin until the 1880s. In 1893, Swiss surgeon Emil Theodor Kocher (1841–1917) predicted that the thyroid might store iodine, but it was Eugene Baumann (1846–1896) who confirmed the presence of iodine as a normal part of the thyroid gland. During this time, scientists began to speculate on the excessive function of the thyroid gland. In 1886, Paul Julius Moebius (1853–1907) suggested that hypersecretion of a ductless gland, like the thyroid, could trigger a disease such as exophthalamic goiter.

While research on iodine thrust forward, Irish physician Robert James Graves (1796–1853) busied himself with experiments on thyroid malfunction. Graves was one of the first physicians to fully describe exophthalmic goiter, a condition that is today called Graves' disease. Graves' disease (see Chapter 10) is characterized by an enlargement of the thyroid gland and a protrusion of the eyeballs (**exophthalmos**). In 1835, Graves' paper *Palpitation of the Heart with Enlargement of the Thyroid Gland* included the following account of exophthalmic goiter:

> A lady, aged 20 became affected with some symptoms which were supposed to be hysterical. . . . After she had been in this nervous state about three months, it was observed that her pulse had become singularly rapid. This rapidity existed apparently without cause and it was constant, the pulse being never under 120 and often much higher. She next complained of weakness on exertion and began to look pale and thin. Thus she continued for a year. . . . It was now observed that her eyes assumed a singular appearance for the eyeballs were apparently enlarged. (Medvei 1993, 142)

Graves, however, never mentioned tremors, which Jean-Martin Charcot (1825–1893) emphatically described in 1856. In 1838, Graves observed his fourth patient with protruding eyes. Around this time, Armand Trousseau (1801–1867) began referring to the disease as "Graves' disease" in his lectures. In 1840, Carl Adolph von Basedow's (1799–1854) celebrated paper *Exophthalmos durch Hypertrophie des Zellgewebes in der Augenhohle* (Exophthalmos through Hypertrophy of the Cellular Tissue in the Orbit) became Europe's most important, and most accurate, account of the disease. Basedow described three women and one man with protruding eyes, enlarged thyroid, and heart palpitations.

Earlier in this chapter, Kocher's important observations regarding the thyroid and iodine uptake are mentioned; however, his most significant achievement was the excision of the thyroid gland for the treatment of goiter in 1876. He was the first surgeon to attempt such a therapy. Seven years later, he reported that total excision of the thyroid gland resulted in a cretinoid pattern in patients, but that leaving part of the gland intact helped reduce the severity of the pathological pattern.

A number of significant discoveries related to diabetes mellitus and the endocrine pancreas occurred in the eighteenth century. In 1815, Michel Eugene Chevreul (1786–1889) concluded that the sugar in the urine of diabetes mellitus patients was glucose. More than three decades later, Carl August Trommer (1806–1879) introduced the first test for glucose in the urine, a test that remained prevalent in Germany until the 1920s. In 1848, Hermann Christian von Fehling (1812–1885) developed the first quantitative test for detecting sugar in urine. Other significant achievements in diabetes during the eighteenth century are outlined in "Notable Eighteenth-Century Findings Regarding Diabetes Mellitus."

Years after Frank distinguished diabetes insipidus from sugar diabetes, Robert Willis (1799–1878) outlined several forms of diabetes insipidus in

Notable Eighteenth-Century Findings Regarding Diabetes Mellitus

- Henry Dewey Noyes (1832–1900) observes eye changes in patients with advanced diabetes mellitus.

- Adolph Christian Jacob Gerhardt (1833–1902) introduces a test for aceto-acetic acid in diabetic urine.

- Prout first describes diabetic coma.

- Adolf Jussmaul (1822–1902) proves that diabetic coma is caused by acetonemia, a condition marked by the presence of acetone bodies in the blood. Ernst Stadelman (1853–?) confirms his finding.

- Oscar Minkowski (1839–1931) identifies the acid in diabetic urine as β-oxybutyric acid.

- Frederick William Pavy (1829–1911) finds a causal link between the degree of sugar in the blood and sugar in the urine.

- In 1876, Wilhelm Ebstein (1836–1912) reports on a diabetes mellitus treatment using sodium salicylate.

his book *Urinary Diseases and Their Treatment*. Prout followed with his publication *An Inquiry into the Nature and Treatment of Diabetes and Calculus,* in which he described excess urea and characterized patients as being with or without polyuria. In 1877, Samuel Jones Gee (1839–1911) provided what is believed to be the first description of familiar nephrogenic diabetes insipidus in his book *Contribution to the History of Polydipsia.* His book gives accounts of eleven patients in a four-generation family who suffered from diabetes insipidus.

Joseph von Mering (1849–1908) and Oscar Minkowski (1858–1931) performed the most important experiment regarding diabetes in the nineteenth century. In 1889, the pair attempted to turn off fat digestion in dogs by surgically removing their pancreas. Their experiment failed, but it had a monumental side effect: the dogs with no pancreas showed signs of diabetes. Minkowski's follow-up tests indicated that the test dogs' urine was 12 percent sugar. Their accidental finding led Minkowski to theorize that the pancreas secreted an "antidiabetic" substance. Today that substance is known as insulin.

Until this time, little was mentioned about the endocrine pancreas. In 1869, German student Paul Langerhans (1847–1888) remarked in his doctoral thesis that the pancreas contained irregularly shaped patches of endocrine tissue. His paper *Beitraege zur mikroskopischen Anatomie der Bauchspeicheldruese* (Contributes to the Microscopical Anatomy of the Pancreas) (1869) documented the pancreatic islets for the first time. In 1893, Gustave-Edouard Laguesse (1861–1927), who had already suspected the cells might harbor important endocrine secrets, termed the tissues "the islets of Langerhans." The normal human pancreas contains about a million islet cells. They are categorized into four different cell types, one of which produces the hormone insulin (see Chapter 5). Langerhans did not, however, report on their function.

The connection between the islets of Langerhans and diabetes was slow to evolve. In 1893, Edouard Hedon (1863–1933) reported that if a small piece of pancreatic tissue was transplanted into the area of the missing pancreas and its circulation remained intact, diabetes would not develop. The long-lived Eugene Lindsay Opie (1873–1971) later established the association between ailing islets of Langerhans and diabetes.

After some quiet, the 1800s saw renewed interest in the pituitary gland (see "Other Highlights of the Nineteenth Century"). For centuries both the anatomy and function of the gland had puzzled scientists. Sir Humphry Davy Rolleston (1862–1944) once said, "The pituitary is complex both in structure and function, and its anatomy is not exactly the same in man as in animals." In 1838, German anatomist Martin Heinrich Rathke (1793–1860) described the embryonic structure where the anterior lobe of the pituitary gland developed. Today this structure is referred to as Rathke's

Other Highlights of the Nineteenth Century

1850	Thomas Blizzard Curling (1811–1888) describes defective cerebral development (cretinism) due to a missing thyroid.
1852	Sir Richard Owen of London (1804–1892) gives one of the first descriptions of the parathyroids in the rhinoceros.
1880	Yvar Victor Sandstrom (1852–1889) offers the first complete description of the parathyroids.
1884	Christian Friedrich Frizsche (1851–1938) and Albrecht Edwin Klebs (1834–1913) publish their clinical and postmortem findings of an acromegalic patient.
1891	Eugene Gley (1857–1930) demonstrates that the parathyroids are essential for life.
1895	George Oliver (1841–1915) and Edward A. Schäfer (1850–1935) develop adrenal extracts that narrow the blood vessels. The principle compound was later named epinephrine.

Sources: V. C. Medvei, *The History of Endocrinology; Grolier Encyclopedia; Encyclopedia Britannica.*

pouch. Adolph H. Hannover (1814–1894) described two kinds of cells in the anterior pituitary lobe; in 1886 several scientists characterized them as nongranular clear cells and granular chromophil cells. The next year, Vincenzo Brigidi published his autopsy results on Ghirlenzoni, an acromegalic actor. His report included the histology of the pituitary tumor. Italy's Andrea Verga (1811–1895) published the first postmortem report of acromegaly in 1864.

French neurologist Pierre Marie (1853–1940) published the first clinical description of acromegaly in 1886; he was the first to give the disease its common name. His published account described the overgrowth of bone tissue in the nose, jaws, fingers, and toes as characteristics of acromegaly. His discovery that the disease stemmed from a pituitary tumor enabled further evolvement of the modern science of endocrinology. Minkowski later reported on a functional relationship between the pituitary tumor and the substance it secreted. Modern surgical treatments to remove pituitary tumors also date back to this time. In 1893, Richard Caton (1842–1926) and Frank Thomas Paul (1851–1941) attempted to surgically relieve brain pressure from a tumor in an acromegaly patient.

Sir Victor Horsely (1857–1916) conducted a number of endocrine experiments, producing solid evidence that **myxedema** and cretinism resulted from an underactive thyroid, although it is noted that some of the symptoms resulted from the removal of the parathyroids. In 1886, he successfully removed the pituitary gland from two dogs. A few years later, Horsely recommended thyroid tissue grafts for myxedematous patients. The idea was eventually modified and successfully carried out by George Redmayne Murray (1865–1939), who injected thyroid extract into his patients.

The year 1855 represented a turning point in the history of endocrinology. Three pioneers, Claude Bernard (1813–1878), Thomas Addison (1793–1860), and Charles-Édouard Brown-Séquard (1817–1894), launched significant ideas during this time. Bernard, regarded by many as the most brilliant physiologist of the time, made significant discoveries concerning the pancreas' role in digestion. In 1856, he noted a white starchy substance in the liver. Further experiments revealed that this complex substance was actually a buildup of sugar that was kept in reserve, then released by the body to keep blood sugar at a constant level. He showed that the liver released sugar into the bloodstream, and he called this process "internal secretion." Bernard's classification distinguished secretions of the liver, namely bile, and the "secretion interne," or the giving off of glucose into the blood. During his lifetime, Bernard analyzed extracts of the thyroid and adrenal glands. The so-called founder of experimental medicine had such an impact on modern medicine that he was given a state funeral, full of pomp and circumstance, an honor ordinarily reserved for princes, statesmen, and soldiers.

London physician Thomas Addison was the first to link certain disease symptoms with pathological changes in one of the endocrine glands, namely the adrenal gland. In 1849, he presented a preliminary account of the disease now named after him: Addison's disease, a rare syndrome defined by destruction of the outer layer of the adrenal glands. The clinical scientist actually stumbled onto the disease in 1849 while searching for the cause of pernicious anemia. The significance of his observation was initially lost. But six years later, at the prodding of a colleague, Addison published his monograph, *On the Constitutional and Local Effects of Disease of the Supra-Renal Capsules,* which evolved into a classic of medicine. It triggered immediate

Thomas Addison. © National Library of Medicine.

experimental studies into the physiological activity of the adrenal glands and, later, separate examination of the cortex and medulla. In 1896 Johns Hopkins Hospital physician Sir William Osler (1849–1919) attempted to treat Addison's patients with an oral adrenal extract. His experiments were cut short when a young female patient died during treatment.

French physiologist Brown-Séquard studied the removal of the adrenal glands. In 1856, he discovered that the adrenal gland is essential for life and eventually proved that so-called internal secretions (later called hormones) helped the body's cells communicate with the nervous system. Brown-Séquard also attempted the first endocrine therapy in 1889. He prescribed extracts from animal testes to treat male aging. His extracts formed the foundation for today's modern cortisone and thyroid hormones.

In the mid-nineteenth century, scientists again revisited the topic of castration and examined its endocrine effects. Medical observers had long known that the testicles were the oldest key to the endocrine system's mysterious ways. In 1849, Arnold Adolph Berthold's (1803–1861) castration experiments resulted in documented endocrine disruption. After castrating a male fowl and performing a single testis transplant, Berthold discovered that its sexual behavior and secondary sex characteristics were completely restored. Berthold never explicitly mentioned that the testis produced a specific substance that maintained the male secondary sexual characteristics. However, scientists later presumed that was the only possible conclusion one could draw from Berthold's experiments.

THE DAWN OF MODERN TIMES: THE TWENTIETH CENTURY

The birth of modern endocrinology begins with British physiologists William Bayliss (1860–1924) and Ernest Starling (1866–1927), codiscoverers of hormones. The pair's meticulous experiments led them to discover the chemical substance that stimulates the secretion of pancreatic digestive juices, the first example of hormonal action. In a 1902 experiment performed on anesthetized dogs, Bayliss and Starling proved that the combination of dilute hydrochloric acid and partially digested food activates a chemical substance in the small intestine. This "secretin," as they termed it, is released into the bloodstream, where it makes contact with the pancreas. Further observations showed that the substance prompted the pancreas to discharge digestive juices through the pancreatic duct and into the intestine. The pair also observed the muscle action responsible for pushing food through the intestines (peristalsis). A few years after their famous experiments, Starling coined the phrase "hormone" to denote secretions released from an endocrine gland into the bloodstream.

The mysteries surrounding the parathyroids began to unravel in the early 1900s. In 1908, American pathologist William MacCallum (1874–1944), and pharmacologist Carl Voegtlin (1879–1960) reported that the muscle twitches (tetany) that occurred after a parathyroidectomy were directly related to the level of calcium in the blood. A year later, MacCallum discovered that injecting patients with calcium salts could alleviate parathyroid deficiency. The earliest report of hyperparathyroidism dates back to 1914, when German pathologist Jakob Erdheim (1874–1937) described enlarged parathyroids in rats with rickets. A decade later, James Bertram Collip (1892–1965) prepared active parathyroid extracts, or parathyrin, and described its successful use in treating a patient with tetany. His discovery came some sixteen years after MacCallum and Voegtlin connected blood calcium deficiencies to tetany. In the early 1930s, American endocrinologist Fuller Albright (1900–1969) performed extensive investigations into calcium metabolism. Albright's far-reaching research in the many fields of endocrinology, including parathyroid disease, bone metabolism, and disturbances of sex hormones, resulted in a number of diseases being named in his honor. Among them are Albright-Butler-Bloomberg disease, a metabolic syndrome marked by dwarfism and other severe developmental abnormalities; Albright's anemia, a form of anemia seen in advanced hyperparathyroidism; and Forbes-Albright syndrome, a disorder marked by galactorrhea and the absence of menstruation due to a pituitary tumor. While at Massachusetts General Hospital in Boston, he and his team published countless studies on primary hyperparathyroidism, including its diagnosis, biochemistry, and prognosis. He described his discoveries in 118 publications, including *The Parathyroid Glands and Metabolic Bone Disease* (1948). In that book, Albright and Edward Conrad Reifenstein (1908–1975) define primary hyperparathyroidism as a "condition where more parathyroid hormone is manufactured than needed."

The pituitary gland's precarious position at the base of the brain delayed experimental investigation. But in 1909, Harvey Cushing (1869–1939) reported that he removed about one-third of a female patient's anterior pituitary lobe and successfully treated an acromegalic patient. Cushing, the leading neurosurgeon of the early twentieth century, performed a great deal of research on the pituitary. He was the first to connect a type of obesity of the face and body to pituitary malfunction, a disease today called Cushing's syndrome (see Chapter 11). The great surgeon introduced numerous innovations to breast and thyroid surgery, and he carried out experimental parathyroid transplants. In his 1909 address to the American Medical Association, Cushing introduced the terms hyper- and hypopituitarism. In summary, Cushing said:

[Hyperpituitarism] expresses itself chiefly as a process of overgrowth—gigantism, when originating in youth, acromegaly when originating in adult life.

[Hypopituitarism] expresses itself chiefly as an excessive, often a rapid, deposition of fat with persistence of infantile sexual characteristics when the process dates from young, and a tendency toward a loss of the acquired signs of adolescence, when it first appears in adult life.

A few years after Cushing's lecture, Philip E. Smith (1884–1970) and Bennet M. Allen (1877–1963), working individually, reported that pituitary gland removal resulted in a slowed growth rate and an underactive thyroid.

The early twentieth century saw an increase in hormone purification, opening doors for new therapies for a host of endocrine diseases. In 1914, American biochemist Edward Calvin Kendall (1886–1972) isolated pure thyroxine from thyroid extracts. It is said that three tons of pig thyroid glands were needed to obtain 33 grams of pure thyroxine. Because the price of production ran higher than the drug's cost, only minute amounts of the substance were available for biological tests until Sir Charles Robert Harington (1897–1972) determined its chemical composition and developed a better method of isolation in 1926.

In 1927, Julius Moses Rogoff (1883–1966) and George Neil Stewart (1860–1930) prepared two different adrenal cortical extracts. But their concoctions still contained traces of dangerous epinephrine, which limited the amount that could be used. Another team, Frank Alexander Hartmann (1883–?) and his colleagues at the University of Buffalo announced they had come up with a way to extract an epinephrine-free cortical substance. Armed with their new substance, the researchers kept completely adrenalectomized cats alive for an average of three weeks, compared to just six days for those without treatment.

During this time, biochemists from the United States and Europe extracted a number of different steroids from the adrenal cortex. In 1933, Kendall and Arthur Grollman independently announced they had crystallized adrenal cortical extracts with hormone activity. Oscar Paul Wintersteiner (1898–?) and Joseph John Pfiffner (1903–?) made similar announcements. Scientists in Berkeley, California, led by Choh Hao Li (1913–1987) reported they had isolated the pure hormone ACTH from sheep pituitary glands. On the other side of the United States, George Sayers (1914–?) and his colleagues at Yale obtained ACTH from pigs.

Swiss chemist Tadeus Reichstein (1897–1996) and his colleagues isolated about twenty-nine hormones and determined their structure and composition. Among these hormones was cortisone, which Reichstein along with Philip S. Hench (1896–1965) discovered had therapeutic benefits for certain arthritis patients. While working at the Mayo Clinic, Hench noted that arthritis pain decreased and in some cases disappeared when the patient was pregnant or had jaundice. This led him to theorize that a biochemical, even hormonal, disturbance was to blame for the disease. After synthesizing cortisone, Hench injected it into a number of rheumatoid arthritis patients with dramatic results. Hench and his colleagues published their results in the *Mayo Clinic Proceedings* in 1949: "In each of the 14 patients . . . [W]ithin a few days there was marked reduction of stiffness of muscles and joints, lessening of the articular aching or pain on motion and tenderness, and significant improvement of articular and muscular function (181–197)." In 1950, Reichstein shared the Nobel Prize for Physiology or Medicine with Hench and Kendall for their independent research on hormones of the adrenal cortex.

Estrone is another important hormone that researchers identified during the twentieth century. In 1929, Edward Doisy (1893–1986) crystallized the female sex hormone estrone in the urine of pregnant women. For a dozen years, he and his colleague Edgar Allen (1892–1943) developed new techniques to expedite research on sex hormones. The scientists also isolated the sex hormones estriol and estradiol. Their two research papers, *Sex Hormones* (1936) and *Sex and Internal Secretions* (1939), are said to have signified the beginning of a new era in female reproductive endocrinology.

Perhaps the most notable achievement of the twentieth century was the discovery of insulin by Canadian physician Frederick Banting (1891–1941) and medical student Charles Best (1899–1978). Since Minkowski and von Mering's discovery in 1889, scientists had made repeated efforts to extract insulin from the pancreas. All attempts failed, however, because digestive enzymes destroyed the insulin molecules the moment the pancreas tissue was crushed. In May 1921, Banting and Best launched extensive experiments to isolate the hormone. They conducted their research in the laboratory of Scottish physiologist J.J.R. Macleod (1876–1935) at the University of Toronto. Banting and Best found that by tying off the pancreatic ducts of dogs they could cause pancreatic inactivity while preserving the islets of Langerhans, which were believed to be the site of insulin production. Banting was so involved in his work that he sold his car to get the money he needed to buy more test animals. When the pair injected a solution extracted from islet cells into dogs without a pancreas, the dogs quickly recovered from their artificially induced diabetes. Banting and Best's research culminated in

Doctors Banting and Best with a dog in their laboratory. © National Library of Medicine.

1922, when they isolated a form of insulin. Their discovery was the first major breakthrough in the understanding of diabetes mellitus; it immediately transformed treatment of the disease. The prognosis of a diabetic patient soon changed from an almost certain death sentence to a long life. In 1923, Banting and Macleod received the Nobel Prize for Physiology or Medicine for their discovery, despite the fact that Macleod never performed the research. Angered that Macleod, and not Best, took home the prize, Banting split his share of the award equally with Best.

Shortly after the landmark discovery of insulin, English chemist and two-time Nobel Prize winner Frederick Sanger (1918–) determined the full chemical structure of insulin. He received the Nobel Prize for Chemistry in 1958 for his research. Sanger spent a decade exposing the structure of the bovine insulin molecule. His achievement was a critical step toward the laboratory synthesis of insulin. In the 1960s, several groups succeeded in producing the hormone synthetically. Sanger's innovative techniques for determining the order in which amino acids are linked in proteins paved the way for the structural determination of many other complex proteins.

While scientists worked on uncovering the molecular structure of hormones, Hans Selye (1907–1982) introduced a term that today is commonplace: stress. In 1936, *Nature* published his landmark paper on stress and its effect on the body. Selye uncovered the damaging effects of stress after injecting ovarian hormones into the glandular system of rats. The hormones stimulated the outer tissue of the adrenal glands. Necrosis of the thymus, ulcers, and finally death resulted. Upon closer examination, Selye discovered that any toxic substance, physical injury, or environmental stress could trigger these effects. Selye eventually extended his theory to humans, demonstrating that a stress-induced breakdown of the hormonal system could lead to heart disease and high blood pressure. In 1950, Seyle published his findings in his major work, *The Physiology and Pathology of Exposure to Stress.* In it, he describes how he stumbled onto the concept that mammals respond to stress or injury with a series of physiological reactions and adapt to them. He described this collection of symptoms as "stress syndrome." Seyle is said to have been influenced by many other scientists, including Walter B. Cannon (1871–1945), an American neurologist and physiologist who discovered sympathin, an adrenaline-like substance released by the endings of certain nerve cells. (Today, sympathin is known to be a mixture of epinephrine and norepinephrine.) Selye authored thirty-three books, including the 1974 international publication *Stress without Distress.*

With many endocrine disease mechanisms now understood, scientific focus turned to diagnosis and treatment. Researchers at the Massachusetts Institute of Technology pondered whether radioactive iodine would allow better studies of thyroid physiology and perhaps even lead to a new agent to diagnose and treat thyroid disease. Scientists Saul Hertz (1905–1950), Arthur Roberts, and Robley D. Evans (1907–1996) were given the task of investigating iodine isotopes. In 1938, the trio reported on the first use of radioactive iodine for thyroid function studies; a few years later, Roberts and Hertz published "Radioactive Iodine in the Study of Thyroid Physiology" (1946) in the *Journal of the American Medical Association*. Therapeutic use of radioactive iodine dramatically reduced the need for thyroid surgery.

In 1960, American medical physicist Rosalyn Yalow (1921–) and her colleague Solomon A. Berson (1918–1972) developed a breakthrough method combining radioactive isotopes with immunology to analyze and diagnose various diseases. They called their method radio-immunoassay or RIA. The advent of RIA gave endocrinologists the power to measure substances that exist in such low concentrations in the body that they are otherwise undetectable. Yalow and Berson used RIA to prove that not all diabetics suffer from a blood insulin deficiency. They hypothesized that an unknown factor blocked the action of insulin in some patients. Their finding enabled endocrinologists to formally recognize the two major types of diabetes: type 1 (insulin-dependent) diabetes and type 2 (non-insulin-dependent) diabetes.

RIA revolutionized the study of endocrinology; the method quickly became an indispensable tool for providing remarkably accurate measurements of hormones and other substances. Today it is used to screen blood for the hepatitis virus and other foreign substances, to identify hormone imbalances in infertile couples, and to determine effective dosage levels of drugs and antibiotics. Yalow and Berson's research partnership lasted twenty-two years until Berson's death in 1972. In 1977, Yalow received the Nobel Prize for Physiology or Medicine for the development of the RIA.

Yalow shared the prize with French-born American physiologist Roger Guillemin (1924–) and Polish-born American endocrinologist Andrew Schally (1926–). Guillemin and Schally, in competition with one another, isolated an amazing number of endocrine hormones, including growth hormone–releasing hormone (GHRH), luteinizing hormone–releasing hormone (LHRH), follicle-stimulating hormone (FSH), and thyrotropin-releasing hormone (TRH). Schally also studied the action of the peptide somatostatin, a substance that inhibits the release of growth hormone. Guillemin proved the old theory that the hypothalamus secretes hormones that control the pituitary gland. He also discovered the endorphins, an important class of proteins involved in the perception of pain.

Growing understanding of the sex hormones opened new doors for contraception in the twentieth century. In 1951, American endocrinologist Gregory Pincus (1903–1967) and his colleagues started investigating how synthesized hormones might inhibit pregnancy. They experimented with manipulating the hormonal environment of the ovum. Pincus's collaboration issued a clinical report detailing the use of oral steroid hormones to suppress ovulation. In 1960, the U.S. Food and Drug Administration granted marketing approval for the first birth control pill.

The latter end of the century saw additional breakthroughs in the treatment of diabetes (see "Other Notable Advances in Twentieth-Century Endocrinology"). In 1968, the oral antidiabetic agent glibenclamide hit the market. The drug alleviated diabetic symptoms by stimulating insulin production and promoting the uptake of sugar in the body's cells. In 1980, the Mayo Clinic proposed intensive insulin therapy to offset the complications of diabetes. During this decade, scientists also introduced the first biosynthetic human insulin (see photo in color insert) and developed the insulin

Other Notable Advances in Twentieth-Century Endocrinology

1916   P. E. Smith and Bennett Allen individually report that removing the pituitary results in a slowed growth rate and underactive thyroid.

1916   A group of endocrinologists form the Association for the Study of Internal Secretions. Today, the group is called the Endocrine Society.

1925   Austrian surgeon, Felix Mandl, is the first to remove a parathyroid tumor from a patient.

1926   Philip E. Smith develops a method to remove the pituitary gland from lab animals.

1933   H. L. Fevold separates gonadotrophic hormones into FSH and LH.

1954   Paul Bell elucidates the molecular weight of ACTH.

1969   Cho Hao Li determines the amino acid sequence of growth hormone. He synthesizes it a year later.

1975   Hyperprolactinemia is first described.

1978   Screening begins for congenital hypothyroidism in the United States, Canada, England, Japan, and other countries.

1986   Octeotride, a somatostatin analogue, is produced for the treatment of acromegaly and other endocrine symptoms.

Sources: Encyclopedia Britannica; The Endocrine Society; Grolier Encyclopedia; V. C. Medvei, The History of Endocrinology, 1993.

pen, an innovative method for self-administering insulin. At the turn of the twenty-first century, researchers launched animal experiments on a diabetic skin patch that administered human insulin and claimed to lower blood sugar levels. Florida scientists announced they had developed a gel patch that could measure blood sugar levels without painful needle sticks. The National Institute of Diabetes and Digestive and Kidney Diseases (NIDDK) launched the Diabetes Prevention Program (DPP) in 1996 with the goal of learning how to prevent or delay type 2 diabetes in people with strong risk factors.

At the close of the century, an alarming theory of environmental endocrine disruption started to emerge. In the 1970s, London scientists began to observe that individuals who worked with biologically active materials complained of fertility problems. Two decades later, evidence continued to mount that environmental exposure to some chemicals could interfere with human and animal endocrine systems. Researchers slowly began to identify a number of assumed endocrine disrupters with structures akin to real hormones. Among them are breakdown products of several pesticides such as DDT and dioxins, a group of toxic chemical byproducts from paper production and incineration.

In the 1990s, the mounting concerns over endocrine disrupters instigated government intervention. Congress ordered the Environmental Protection Agency (EPA) to establish a screening program, citing "a growing concern over the effect of pesticides and other substances on human endocrine systems and their ability to increase the likelihood of disease" (The Why Files). In 1996, the EPA's Office of Research and Development labeled endocrine disruption as a top research priority and developed an approach to address some of the uncertainties. The Endocrine Disrupter Research Planning Act of 1996 called for more coordinated studies to better assess the health risks of endocrine disrupters. Today, the hypothesis of endocrine disrupters remains a subject of sore debate. While evidence suggests that endocrine systems of certain fish and wildlife are affected by chemical contaminants, the relationship between endocrine disease in humans and environmental contaminants is yet to be fully understood. The subject will undoubtedly be a topic of research for years to come.

The Future

Every day, scientists are enhancing their knowledge of endocrine system functions and learning how to better treat related conditions. Advances in genetics, tissue engineering, molecular biology, and other disciplines are helping reinvent the field of endocrinology in the new millennium.

One of the hottest topics of research in the twenty-first century is the endocrine aspects of obesity. Obesity is a monumental healthcare problem that continues to grow in epidemic proportions in the United States. According to the Office of the U.S. Surgeon General, the risks of being overweight or obese may soon cause as much disease and death as cigarette smoking. Recent studies indicate that about 15 percent of American kids aged 6 through 19 are severely overweight or obese, putting them at increased risk of diseases like diabetes.

At the turn of the twenty-first century, endocrinologists were paying close attention to a newly defined disorder called metabolic syndrome, a clustering of medical conditions including obesity. The syndrome was first described in the Third Report of the National Cholesterol Education Program (NCEP) Expert Panel on Detection, Evaluation, and Treatment of High Blood Cholesterol in Adults, released in 2001. More than one in five Americans has metabolic syndrome (also known as Syndrome X). Experts say this high prevalence underscores an urgent need to develop wide-ranging efforts directed at reducing, if not eliminating, the obesity epidemic. Many researchers theorize that insulin resistance is a key player in the development of this syndrome, although there is no complete consensus.

As science takes off in this new millennium, research on metabolic syndrome is sure to follow. Already, endocrinologists are unraveling the detailed mechanisms behind appetite control and body weight. In the August 2002 *Journal of Clinical Endocrinology & Metabolism* (*JCEM*), researchers at the University of California, Los Angeles, School of Medicine announced that insulin levels have a direct impact on a recently identified hormone called ghrelin, which stimulates hunger. Their finding will help endocrinologists move closer to understanding the mechanisms behind, and possible treatments for, obesity (see photo in color insert). A second independent study on ghrelin, also published in the *JCEM,* found that a common **polymorphism** (SNP247) of the ghrelin gene correlates to an increase in body mass index and reduced insulin secretion. The London researchers noted that SNP247 was associated with an earlier onset of childhood obesity. Advancements in the understanding of the hormonal control of food intake and energy expenditure can be used to design more effective obesity treatments.

The success of the Human Genome Project greatly accelerated research into identifying disease-related genes. Advances in genetic research are having a profound effect on our understanding of the two most common endocrine disorders: autoimmune thyroid disease (see Chapter 10) and type 2 diabetes (see Chapter 13). Today scientists are hunting for the elusive genes that may be responsible for type 1 or type 2 diabetes. A report in the July 2003 edition of the *Proceedings of the National Academy of Sciences* detailed the successes of an ambitious and exhaustive genetic study led by investigators at the Joslin Diabetes Center and the Children's Hospital Boston Informatics Program. The researchers pinpointed a family of genes that are heavily involved in the development of type 2 diabetes. Study investigator Mary-Elizabeth Patti said, "Knowing which genes are turned on or off in people *before* they develop diabetes is a key piece of information needed to solve the puzzle of diabetes, and to identify new ways to treat and prevent this devastating disease."

Other scientists have successfully revealed some genetic markers for type 1 diabetes. A massive amount of research funding has been allocated to identify the genetic and environmental contributors to type 1 diabetes. The Special Statutory Funding Program for Type 1 Diabetes provides a total of $1.14 billion in research funds from fiscal year 1998 through fiscal year 2008. In addition to identifying genetic and environmental ties, the research projects will address the following broad goals: to prevent or reverse type 1 diabetes, to develop cell replacement therapy, to prevent or reduce hypoglycemia in type 1 diabetes, to prevent or reduce the complications of type 1 diabetes, and to attract new scientists to research type 1 diabetes. The number of diabetes-related clinical trials continues to expand. The *Type 1 Diabetes TrialNet* is a collaborative network of clinics, labs, and experts that are focused on testing new approaches to understanding, preventing, and treating type 1 diabetes. Research is expected to focus on stopping or delaying the immune destruction of the insulin-producing beta cells of the pancreas in people at risk for type 1 diabetes, an autoimmune disease. Studies have already proven that the immune system continues to destroy beta cells even after a type 1 diabetes diagnosis, making blood glucose harder and harder to control for many people.

New therapies for diabetes are also being researched. The *New England Journal of Medicine* (*NEJM*) reported in 2002 that an experimental drug called anti-CD3 antibody (anti-CD3 mAb) may delay the typical decline in insulin production in people with newly diagnosed type 1 diabetes. Research on this compound is ongoing. In July 2003, endocrine researchers at Hoffman-La Roche announced that they had discovered a family of molecules that impact the key gene involved in blood sugar regulation. Experts are optimistic that the finding will help them develop new ways to control blood sugar. Researchers at the NIDDK and the University of Texas Southwestern Medical Center are using the hormone leptin to treat patients suffering from **lipodystrophy,** a rare and difficult-to-treat metabolic disorder that shares some of the characteristics of typical type 2 diabetes. The finding, which is the first example of leptin's use to treat lipodystrophy, also appeared in the *NEJM*. Amylin Pharmaceuticals is conducting preliminary research on a new, synthetic biological compound called exanatide. Early studies have shown that, when added to existing antidiabetes treatment, exanatide improves blood glucose levels in type 2 diabetics. The report, published in the journal *Diabetes Care,* said that the compound exanatide reduced blood levels of fructosamine, which is a measure of intermediate (two weeks) glucose control, and hemoglobin A1c (HbA1C), which is a measure of longer-term (three months) glucose control. Swiss drug giant Roche is conducting animal trials on a new class of drugs that shows promise in attacking type 2 diabetes on more than one front. Their experimental drug, RO-28-1675, is a glucokinase (GK) enzyme activator; it acts on GK, an enzyme that diabetes researchers have known about for years. RO-28-1675 stimulates the pancreas to release more insulin and keeps the liver from producing too much glucose.

Several new antidiabetic drugs made it through discovery and research and into consumers' hands at the commencement of the new millennium. A variety of oral diabetes medications are now available, allowing diabetics to control blood glucose levels without insulin injections. Studies are underway to determine how best to use these drugs to manage type 2 diabetes. Scientists also are investigating strategies for weight loss in people with type 2 diabetes.

Endocrinologists say transplantation of the pancreas or insulin-producing beta cells of the pancreas offers the best hope for a cure for type 1 diabetes. In recent years, doctors have reported some successful pancreatic transplants. Today, scientists are working to develop less toxic drugs and better methods of transplanting beta cells to prevent rejection. Bioengineers are working to create artificial beta cells that secrete insulin in response to increased glucose levels in the blood. Recently, researchers at the University of Alberta in Edmonton, Canada, announced that islet transplantation in seven patients with type 1 diabetes allowed the recipients to remain free of insulin injections for a long period of time after the procedure.

Endocrinologists say tissue engineering could present exciting new courses of action for treating a variety of endocrine maladies. Perhaps the next great frontier in diabetes research, then, lies in the hands of stem cell researchers. Diseases such as diabetes, which have no cure, are a focus of stem cell research. Stem cells are essentially "blank" cells that can be programmed to grow into a number of different cell types, such as insulin-producing islets. Experts argue that transplantation of islets produced from stem cells could potentially cure patients with type 1 diabetes. Several recent studies emphasize that stem cell research could provide much-needed therapies for type 1 diabetes. Researchers have already created insulin-producing islets in embryonic stem cells from mice as well as humans. According to the Juvenile Diabetes Research Foundation, animal studies performed in Canada suggest that transplanted adult stem cells taken from bone marrow can prompt the recipient's pancreatic tissue to heal itself, restoring normal insulin production and reversing diabetes-associated symptoms. The findings, published in the journal *Nature Biotechnology,* also revealed a striking observation that the individual stem cells did not directly heal the pancreas; instead they appeared to enlist the transplant recipient's own cells to multiply and create insulin in response to glucose. The exact mechanism remains a subject of study.

The way certain forms of diabetes are diagnosed could change in the twenty-first century. Massachusetts General Hospital researcher Ravi Thadhani is studying whether a blood test given during the first trimester of pregnancy may help doctors predict which women will develop diabetes. Currently, gestational diabetes is diagnosed in the third trimester with a blood sugar test (see Chapter 13). For the study, Thadhani's team measured blood levels of a protein called sex hormone–binding globulin (SHBG). Results published in the *American Journal of Obstetrics and Gynecology* revealed that women who developed gestational diabetes exhibited lower SHBG levels than women who remained diabetes-free throughout their pregnancy. The study authors say more research is needed to confirm the findings.

Diagnosing other endocrine diseases is expected to get easier as researchers progress into the future. Thyroid function tests have already improved considerably in the last two decades, making early diagnosis much easier. Improved techniques for distinguishing ACTH-dependent forms of

Cushing's syndrome from adrenal tumors are helping doctors better decide the proper course of treatment. In April 2002, SecreFlo, a synthetic, injected form of the hormone secretin, was approved as a diagnostic tool for identifying pancreatic dysfunction and the presence of a potentially cancerous gastrin-secreting tumor called a gastrinoma.

Drug delivery for a multitude of endocrine disorders is becoming less invasive and more convenient. At the turn of the century, Pharmacia & Upjohn introduced the Genotropin MiniQuick Growth Hormone Delivery Device, a prefilled, premeasured, single-use device for adults who must take daily growth hormone replacement therapy. Medtronic MiniMed, Inc., and Becton Dickinson received approval from the FDA for their diabetic combination device that integrates a glucose meter and insulin pump with a dose calculator. A number of new endocrine-related drugs has recently been introduced to the market. In 2003, U.S. federal regulators approved Eli Lilly's hormone, Humatrope, a synthetic form of growth hormone that is used to boost height in short children who are otherwise healthy and who have no growth hormone deficiency. The FDA approved the treatment for the shortest 1.2 percent of children. For 10-year-old boys and girls, for example, that would be a height of less than 4 feet 1 inch. The FDA also approved Somavert (pegvisomant) for acromegalic patients who have had an inadequate response to existing therapies. The drug is the first in a new class of drugs called growth hormone receptor antagonists. In studies, pegvisomant normalized concentrations of IGF-I in more than 90 percent of patients by blocking the effects of growth hormone. As new therapies are introduced, however, new drug warnings are emerging as well. In the summer of 2003, the world's largest drugmaker, Pfizer Inc., alerted doctors of several deaths linked to the use of its Genotropin growth hormone. All of the deaths occurred in children with a rare genetic disease called Prader-Willi syndrome.

Pioneering genetic research is helping further the understanding of other endocrine diseases. Investigations are underway to determine the specific cause and location of the gene responsible for dwarfism. Medical theory suggests that disruptions on different areas of chromosomes 3 and 7 play a role in dwarfism. Scientists are working on isolating those defects. Other researchers are looking into the causes of benign endocrine tumors, such as those responsible for most cases of Cushing's syndrome. Specific gene defects have been identified in a few pituitary adenomas. As research progresses, evidence continues to mount that tumor formation is a multistep process. Understanding the basis of Cushing's syndrome will yield new approaches to therapy.

Several new subspecialties of endocrinology were slowly emerging in the early twenty-first century. Cosmetic endocrinology, currently practiced as a part of dermatology, is expected to become a significant area of research and practice. Cardiovascular endocrinology, a little-known area of medicine in the 1900s, is slowly becoming an accepted endocrine subspecialty. The eighty-fifth annual meeting of the Endocrine Society, held in June 2003, focused on this burgeoning field. One of the hottest topics in this arena is glucocorticoids. These popular anti-inflammatory steroid hormones are used to treat a host of health problems including asthma, arthritis, and inflammatory bowel disease. However, scientists are just starting to note a multitude of dangerous side effects associated with glucocorticoids, including increased rates of obesity and high blood pressure. A study presented at the Endocrine Society meeting revealed that people on steroids had a 32 percent chance of experiencing a cardiovascular problem in ten years of follow-up, compared with a 19 percent risk in those not taking steroids. Conference researchers also presented new findings regarding primary aldosteronism, a form of hypertension that occurs when one or both adrenal glands release too much of the hormone aldosterone. Australian scientists discovered that this disorder is much more common than was previously believed. Other researchers have reported similar findings.

Geriatric endocrinology is another emerging field. Just a few centuries ago, humans weren't expected to live past 50 or 60 years. Longer life spans have resulted in a barrage of new health problems. Hormone production is profoundly affected by aging. Scientists are beginning to note that age-related hormonal changes pose a significant threat to the human endocrine system. As knowledge of the endocrine system continues to expand, scientists hope to learn how to extend and improve the action profile of hormones.

So what lies ahead in the field of endocrinology? Experts predict that they will be better able to define and correct the defective mechanisms that lead to the development of endocrine tumors. Improvements in imaging methods will allow for more accurate examination of endocrine cells. Some scientists theorize that, in the near future, the world will see an improvement in site-specific hormone replacement therapy. Others foresee the ability to predict and possibly prevent many more thyroid diseases.

The next generation of endocrinologists, however, could be a rare breed. Numerous studies suggest that endocrinology is a discipline in danger. In April 2003, a multiorganizational study found that endocrinologists are already in short supply and warned that the shortage will grow worse in the coming years. A shortage of specialists combined with an increasing number of endocrine patients, especially those suffering from obesity, could spell disaster. For this reason, the specialized expertise of the endocrinologist is becoming increasingly more important, noted Robert Rizza, lead author of the study. "As people in the United States become more overweight and diabetes continues to grow in prevalence, the specialized expertise of endocrinologists will be increasingly important," said Rizza. "We need to take steps to stop the ongoing decline in the number of endocrinologists in training and find a way to expand the number of endocrinologists in practice in the years ahead." The study, conducted by the Lewin Group, predicted that the already-long delays patients face in getting appointments with an endocrinologist will worsen by 2020 unless the number of new endocrinologists entering the field skyrockets. According to the study, today's supply of endocrinologists is 12 percent less than demand. From 1995 to 1999, the number of endocrinologists entering practice plunged 15 percent. Experts caution that increasing pressures on health care practices could fuel an increase in retirement rates among practicing endocrinologists, worsening the shortage. Robert Vigersky, the Endocrine Society's chairman of clinical affairs, emphasized the critical need for new ways to meet the demand so that patients can receive adequate care.

Diseases of the Pituitary Gland

The pituitary gland serves as the body's control center for long-term growth, kidney and thyroid function, and reproductive capabilities (see Chapter 2). Pituitary disorders can be extremely tricky to diagnose because the wide spectrum of hormonal and neurological symptoms can mimic those associated with other diseases. This chapter is divided into three sections: hypopituitarism (underactive), hyperpituitrism (overactive), and pituitary tumors.

HYPOPITUITARISM

Hypopituitarism is a general term that refers to an underactive pituitary gland. The disorder derives its name, in part, from the Greek prefix *hypo,* meaning under. It occurs when the pituitary gland produces insufficient amounts of one or more hormones. Hypopituitarism can result from damage to the pituitary gland or hypothalamus, head injury, tumor, radiation therapy, stroke, inflammation, and infection. Occasionally, rare immune system or metabolic diseases such as **sarcoidosis, histiocytosis X,** and **hemochromatosis** can cause the deficiency. Postpartum hypopituitarism, or Sheehan's syndrome, may occur after a severe uterine hemorrhage during childbirth. The massive blood loss causes tissue death in the pituitary gland. When all hormones released by the anterior pituitary gland fail, the disorder is referred to as **panhypopituitarism,** or complete pituitary failure.

A patient's symptoms depend on the severity of the deficiency and the types of hormones affected. Pituitary dwarfism and diabetes insipidus, two common disorders caused by an underactive pituitary gland, are explained

TABLE 10.1. Underactive Pituitary Hormone (Hypopituitarism) Disorders

Hormone	Symptoms
Growth hormone (GH)	A deficiency in this hormone can lead to stunted growth and pituitary dwarfism. (See "Pituitary Dwarfism" in this chapter.)
Gonadotrophins: luteinizing hormone (LH) and follicle-stimulating hormone (FSH)	Abnormally low levels of these two hormones can lead to infertility in both men and women. In men, the deficiency can result in impotence, decreased libido, shriveling of the testes, low sperm count, loss of body and facial hair, weakness, fatigue, and anemia. Premenopausal women may experience vaginal dryness, amennorhea (the absence or stopping of a menstrual cycle), decreased libido, and osteoporosis. Treatment consists of estrogen replacement therapy (women) and testosterone injections (men).
Thyroid-stimulating hormone (TSH)	When the pituitary gland fails to produce adequate levels of TSH, the thyroid gland operates with less energy. The result is hypothyroidism. (See Chapter 11.)
Adrenocorticotropic hormone (ACTH)	Insufficient amounts of this pituitary hormone, which regulates adrenal function, cause the levels of cortisol and other steroid hormones in the blood to plummet. Symptoms include weight loss, lack of appetite, weakness, nausea, vomiting, and low blood pressure. Shortage of ACTH resulting in cortisol deficiency is the most dangerous of all hormone deficiencies. (See Chapter 12.)
Prolactin	Women with this rare deficiency are unable to produce breast milk after childbirth. There are no known symptoms in men.
Antidiuretic Hormone	A deficiency in this hormone can cause diabetes insipidus. (See "Diabetes Insipidus" in this chapter.)

in further detail in this chapter. In general, a person with hypopituitarism may complain of fatigue, sensitivity to cold, weakness, decreased appetite, weight loss, abdominal pain, low blood pressure, and headache. Table 10.1 shows the disorders and symptoms associated with reduced production of specific pituitary hormones.

Pituitary Dwarfism

An exaggerated slow-growth pattern in childhood can often be blamed on the underproduction of growth hormone. The resulting condition, pituitary dwarfism, involves abnormally short stature with normal body proportions (see photo). It is characterized by slow growth before age 5 and an absent or slowed height increase (below the fifth percentile on a standardized growth chart). Pituitary dwarfism tends to run in families.

Children with this disorder typically progress normally for the first two or three years of life, then suddenly fall behind the normal growth curve.

Sexual development may be absent or delayed in adolescents. Other symptoms include headache, excessive thirst, and increased urine volume.

If pituitary dwarfism is suspected, the physician will order blood tests to check growth hormone secretions over a period of eight to twelve hours. (Normal growth hormone secretion fluctuates throughout the day, rising after bedtime.)

Treatment consists of human growth hormone injections given daily or several times a week. When therapy is administered before growth plates have fused, most children reach a normal adult height.

Diabetes Insipidus (DI)

This uncommon pituitary disorder occurs when the body produces too little antidiuretic hormone (ADH). Diabetes insipidus is a misnomer of sorts, as it is unrelated to type 1 and type 2 diabetes (see Chapter 12), which are associated with insulin resistance or deficiency. A shortage of antidiuretic hormone causes the kidneys to lose control of their water balance. ADH

Studio portrait of Lilli Walton, a dwarf. Courtesy of the Library of Congress.

normally directs the kidneys to concentrate urine by returning excess water to the bloodstream and therefore make less urine. Too little ADH causes an imbalance in the system, disrupting the kidneys' handling of fluids and causing excessive excretion of dilute urine. Patients become dehydrated and may feel the need to drink large amounts of water. They are likely to urinate frequently, even at night, and may suffer from bedwetting. Children with DI may also have fever, vomiting, or diarrhea. When DI results from damage to, or a tumor in, the pituitary gland, it is referred to as central DI. Because symptoms closely parallel those associated with type 1 and type 2 diabetes, a physician may first suspect an insulin problem. Therefore, a series of tests are performed to rule out traditional diabetes and diagnose DI. Urinalysis allows physicians to note the dilution of urine. The urine of a person with DI will be less concentrated and exhibit lower salt and waste concentrations. A fluid deprivation test helps determine whether DI is caused by a pituitary defect or kidney malfunction. This test measures changes in body weight, urine output, and urine composition when fluids are withheld. In some patients, a magnetic resonance image (MRI) of the brain is used to diagnose central DI.

DI is treated effectively with a synthetic hormone called desmopressin, which can be taken via injection, nasal spray, or pill. Patients on desmopressin should drink fluids or water only when they are thirsty, because the drug prevents water excretion. If a pituitary tumor is the underlying cause, surgery may be needed (see "Pituitary Tumors" in this chapter).

HYPERPITUITARISM

Hyperpituitarism is a vague catchall for diseases that evolve when the pituitary gland goes into overdrive and produces excessive amounts of hormones. Acromegaly and gigantism are two major disorders that can occur as a result of an overactive pituitary gland.

Acromegaly

When the pituitary gland continually produces too much growth hormone (GH) after bone growth ceases, the condition is called **acromegaly**. The disease derives its name from the Greek words *acro* for end or extremities and *megaly* meaning enlargement.

Acromegaly typically strikes adults between the ages of 30 and 50. (If it develops in childhood, the disease is called gigantism. See separate entry in this chapter.) It is quite rare; doctors estimate that only three out of a million people are diagnosed each year. However, because acromegaly is frequently misdiagnosed, these numbers may actually be higher.

Swelling of the hands and feet are among the first symptoms; a patient might note that his or her rings are too tight (see photo in color insert). Eventually, excessive amounts of growth hormone cause the skeleton and internal organs, including the liver, spleen, kidneys, and heart, to grow abnormally large. Bony changes slowly distort facial features; eventually the nose becomes bigger, the brow and lower jaw protrude, and the spacing of teeth increases.

Patients may also exhibit thick, coarse, oily skin; skin tags, and an enlargement of the lips, nose, and tongue. Enlargement of the sinuses and vocal cords can cause deepening of the voice and snoring. Other symptoms include excessive sweating and skin odor, fatigue and weakness, headaches, visual problems, changes in a woman's menstrual cycle, and impotence in men.

Complications from acromegaly range from carpal tunnel syndrome (triggered when thick tissue traps nerves) to arthritis, diabetes mellitus (see

Chapter 12), and high blood pressure. Patients are at increased risk of cardiovascular disease and colon polyps, which can lead to cancer.

The extremely slow progression of the disease often delays diagnosis considerably. Patients may have the disease for more than a decade before seeking help. Early diagnosis is critical, because persons with untreated acromegaly have a shortened life expectancy.

The most reliable method of confirming an acromegaly diagnosis is the oral glucose tolerance test. Ingestion of glucose sugar lowers blood GH levels in healthy people, but the reduction does not occur in patients who overproduce GH. Physicians may also order computed tomography (CT) scans or MRI scans of the pituitary to scout for any tumors. If scans fail to detect a pituitary tumor, the physician will look for tumors in the chest, abdomen, or pelvis as the cause for excess GH.

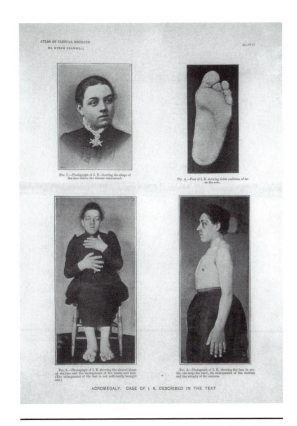

Studies in acromegaly. © National Library of Medicine.

Untreated, acromegaly can lead to severe illness and death. Treatment options include brain surgery, medications, and radiation therapy. In the majority of acromegaly patients, the overproduction of GH results from a benign pituitary tumor (see "Pituitary Tumors" in this chapter). In those cases, surgery is the preferred option, because tumor removal rapidly relieves brain pressure and lowers GH levels. Successful removal of a pituitary tumor reduces soft tissue swelling within a few days. Surgery is most successful in patients with blood GH levels below 40 ng/ml before the operation and with pituitary tumors no larger than 10 mm in diameter. If surgery fails to normalize hormone levels, patients may begin a course of drug therapy. Bromocriptine (Parlodel) and octreotide (Sandostatin) are older medications that reduce both GH secretion and tumor size. In March 2003, the U.S. Food and Drug Administration (FDA) approved pegvisomant (Somavert) for the treatment of acromegaly. In clinical studies, pegvisomant, the first in a new class of drugs called growth hormone receptor antagonists, normalized concentrations of IGF-I in more than 90 percent of patients by blocking the effects of growth hormone. The drug is given by injection. Radiation therapy is generally reserved for patients whose tumors could not be completely removed by surgery or who have health problems that make surgery impossible.

Gigantism

When the pituitary gland releases too much growth hormone in children who are still growing, the result is gigantism.

Gigantism is very rare. The disease usually becomes apparent in early adolescence, when the child is notably larger than his or her peers. Bones grow at an extremely rapid pace creating an extremely tall stature. Robert Wadlow, the tallest giant on record, stood 8 feet 11 inches tall at age 22. (See "Robert Wadlow, the Gentle Giant.") Patients with gigantism have a prominent jawline and disproportionately large hands and feet with thick fingers and toes. Muscles and organs also become abnormally large. The bulk, however, does not translate into strength. Gigantism patients have very weak muscles, putting them at risk for arthritis and heart disease. The disease may delay puberty. Females often have irregular menstrual cycles and may secrete breast milk when they are not nursing. Other symptoms include weakness, headache, double vision, or difficulty with peripheral vision.

Most of the time, gigantism is caused by a pituitary tumor (see separate entry in this chapter). In pituitary tumors with clear-cut borders, surgery is the treatment of choice and is successful in about 80 percent of cases. When surgeons are unable to remove the entire tumor, patients are prescribed GH

Robert Wadlow, the Gentle Giant

When Robert Wadlow was born on February 22, 1918, he weighed in at a relatively average 8 pounds, 6 ounces. He appeared to be a healthy, happy, baby boy. But within months of his birth, Wadlow's parents began to realize that their baby was anything but normal. By the time he was six months old, he weighed 30 pounds. At 18 months, he topped the scales at 62 pounds. By age 8, when the average boy stands a little over 4 feet tall, Wadlow had grown to 6 feet 2 inches tall and weighed 195 pounds. Doctors attributed his extraordinary growth to an overactive pituitary gland, but they could do little to treat him. Therapies available to combat the problem were not yet available in the 1920s.

At age 18, Wadlow entered the *Guinness Book of Records* as the world's tallest man. He measured 8 feet 4 inches. And still, he continued to grow. When his astonishing growth spurt finally stopped, he stood at a whopping 8 feet 11 inches. Finding clothing to fit a man of that size wasn't easy. His size 37 shoes cost $100 a pair, which in the 1930s was a princely sum.

Despite his enormous stature, Wadlow had a sweet nature, which earned him the nickname, "Gentle Giant." Wadlow was only 22 years old when he died in 1940 from an infected foot blister. His 1,000-pound casket required the efforts of a dozen pallbearers.

reduction medications. Radiation therapy is used only as a last resort when surgery and medication fail.

PITUITARY TUMORS

One in five people develops pituitary tumors; most are benign. Despite continued research, the cause of these tumors remains unknown. Most arise without genetic bias; very few are inherited. Research has not discovered why these tumors are so common. In the past, pituitary tumors were not included in statistics collected by state cancer registries and the National Cancer Institute. However, the Benign Brain Tumor Cancer Registries Amendment Act, passed by the U.S. Congress in October 2002, now mandates that hospitals, clinics, and doctors report pituitary tumor incidence rates when collecting data for cancer registries.

Pituitary tumors can cause a number of major endocrine disorders including acromegaly, gigantism, and Cushing's disease (see Chapter 12). Very large tumors can destroy the pituitary gland, leading to hypopituitarism.

Doctors classify pituitary tumors according to size. Microadenomas are tumors that are less than 0.39 inch (1 centimeter) in diameter. Almost all pituitary tumors are of this variety. These tumors never spread outside the skull and usually remain confined to the gland itself. Larger tumors are called macroadenomas. If the tumor grows larger than an inch, it can put pressure on the optic nerve (leading to visual disturbances and sometimes blindness), the brain (impacting function), and the pituitary gland itself (affecting hormone production). Macroadenomas can also invade the sinuses.

Most pituitary tumors secrete hormones, such as growth hormone or prolactin. Tumors are named for the specific cell type affected. For example, if a tumor originates in a cell that produces the hormone prolactin, the patient develops a prolactinoma (see Table 10.2). Knowing what kind of hormone an adenoma produces dictates which tests are used for diagnosis, treatment, and survival outlook. Pituitary tumors that do not produce hormones are referred to as nonfunctional tumors. Depending on their size, these nonhormone-secreting tumors can cause headaches, visual field defects, and hypopituitarism.

Prognosis depends on tumor type and how far it has spread, as well as the patient's age and overall health. Patients typically do well when the tumor is diagnosed early. Treatment follows three courses of action: surgery to remove the tumor, radiation therapy, and medication to shrink and sometimes eradicate the tumor and/or block hormone secretions.

Prolactinoma

Also called lactotroph adenoma, a prolactinoma is the most common type of pituitary tumor. This benign mass secretes excessive amounts of the hor-

TABLE 10.2. Types of Hormone Secreting Pituitary Adenomas

Lactotroph adenoma	These prolactin-producing tumors account for approximately 30 percent of pituitary adenomas.
Somatotroph adenoma	This kind of tumor secretes growth hormone; 15–20 percent of all pituitary adenomas are of this variety.
Corticotroph adenoma	An ACTH-producing tumor, it represents about 10–15 percent of pituitary adenomas.
Gonadotroph adenoma	An FSH- and/or LH-secreting tumor, these adenomas account for a small percentage of pituitary tumors.
Thyrotroph adenoma	This uncommon tumor produces thyroid-stimulating hormone.

Source: American Cancer Society.

mone prolactin, which normally stimulates the breasts to produce milk during pregnancy. The disease progresses silently in about 5–10 percent of the adult population.

Women with a prolactin-producing tumor may secrete breast milk outside of pregnancy, a condition called galactorrhea. They may also experience irregular menstrual cycles or miss their period altogether. In men, impotence is the most common symptom. Infertility can result in both men and women. Too much prolactin in the blood creates a drop in sex hormones, triggering bone loss (osteoporosis) in both sexes. Diagnosis of prolactinoma is confirmed with blood work and medical imaging tests, such as MRI and CT scans. Physicians must first rule out other causes for prolactin elevations, such as pregnancy, stress, hypothyroidism (see Chapter 11), kidney failure, liver failure, and medication side effects.

Craniopharyngioma

Craniopharyngiomas are not really pituitary tumors, but their positioning near the pituitary gland permits their inclusion in most medical literature on pituitary tumors. About 200 Americans are diagnosed with a craniopharyngioma each year. Most patients are children, although craniopharyngiomas are sometimes found in adults ages 50 and over.

Craniopharyngiomas are nonfunctional tumors; they do not secrete hormones. But their ability to compress the pituitary gland can disrupt normal hormone production. The tumors are not cancerous, but they can create a host of neurological problems, such as vision disturbances. Children with these tumors experience delayed growth and often become overweight. Other symptoms include **amennorhea** (the absence or stopping of a menstrual cycle), loss of sexual drive, increased sensitivity to cold, fatigue, constipation, dry skin, nausea, low blood pressure, and depression. If the pituitary stalk is compressed, diabetes insipidus or galactorrhea may result.

Craniopharyngiomas typically go undiagnosed until they press on the surrounding brain structure. Late discovery means the tumors are frequently quite large (over 3 cm) when detected. Diagnosis involves a thorough neurological examination, medical imaging tests such as an MRI or CT scan, and blood work to look for hormone imbalances. Brain surgery is usually the main treatment for craniopharyngioma.

Other Pituitary Tumors

There are several other types of pituitary tumors. Teratomas, germinomas, and choriocarcinomas are all uncommon tumors that occur most often in children or young adults. Pituitary gangiocytomas and Rathke's cleft cysts (RCC) are uncommon tumors that are usually found in adults. RCCs occur when the pouch that eventually forms the pituitary gland fails to close during fetal development. The malfunction creates a cleft that lies within the pituitary gland. Occasionally, the cleft gives rise to a large cyst. RCCs are usually asymptomatic but may lead to visual disturbances, symptoms of pituitary dysfunction, and headaches.

Diseases of the Thyroid and Parathyroid Glands

A poorly functioning thyroid or parathyroid gland can trigger a series of health problems. Unlike other hormones that affect only one organ, discrepancies in thyroid hormone levels can wreak havoc on nearly all tissues of the body.

According to the American Thyroid Association, diseases of this type are very common; more than 13 million Americans have some kind of thyroid disorder. Most disorders are classified as hormone deficiency syndromes (hypofunction) or endocrine excess syndromes (hyperfunction) (see illustration).

HYPOTHYROIDISM

Hypothyroidism is, by far, the most common thyroid disease. It occurs when the thyroid gland fails to produce enough of the hormones needed to maintain the body's normal metabolism. Usually, the condition is primary in nature, meaning that it results from failure of the thyroid gland itself. Abnormally low levels of thyroid hormones in the blood characterize this "underactive" disorder. The disease progresses slowly over months or even years. It can occur at any age and in either sex, although women are most commonly affected.

Patients with the disease typically feel mentally and physically sluggish, because the shortage of thyroid hormones slows down the body's normal function rate (**basal metabolic rate [BMR]**). Initial symptoms may consist only of an inability to stay warm in cool or cold temperatures and constant

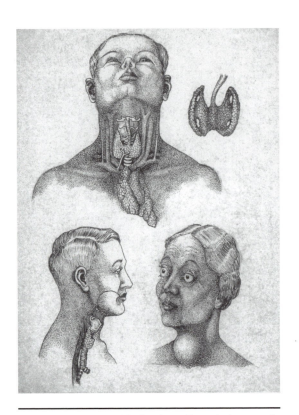

Surgical illustration by Herbet Louis Treusch showing abnormal thyroid growth in male and female patients. © National Library of Medicine.

fatigue. The patient's decreased metabolism can trigger modest weight gain. Other symptoms of adult hypothyroidism are outlined in "Common Symptoms of Hypothyroidism in Adults."

Hypothyroidism presents differently in children than adults; in young patients it has more serious consequences. One out of every 4,000 babies born in the United States has hypothyroidism; girls are twice as often affected than boys (The MAGIC Foundation, 2000). The condition is more common in children with diabetes, rheumatoid arthritis, and Down's syndrome. Most affected infants have only a mild hormone deficiency and, as a result, present few or no symptoms. However, infants who are profoundly hypothyroid have a puffy face and an enlarged tongue that protrudes from a usually opened mouth. Other symptoms include respiratory distress syndrome, poor eating habits (infants may choke frequently), hoarse cry, constipation, jaundice, **umbilical hernia,** and retarded bone age. Left untreated, congenital hypothyroidism can lead to mental retardation and dwarfism, a condition called **cretinism.** When caught early, prognosis is excellent. Today, hospitals in North America routinely screen for hypothyroidism two days after birth.

Advanced hypothyroidism, or myxedema coma, can be life-threatening. This rare condition is marked by intense drowsiness and intolerance to cold, followed by profound lethargy and unconsciousness. Other signs of myxedema coma include unresponsiveness, decreased breathing, low blood pressure, low blood sugar, and below normal temperature. This condition requires immediate medical treatment.

Common Symptoms of Hypothyroidism in Adults

- Weakness, fatigue, lethargy

- Slowed heart rate

- Mental slowness

- Cold intolerance

- Dry skin and hair

- Prolonged reflex times

- Decreased sweating

- Weight gain

- Thick tongue

- Goiter (in some patients)

- Slow speech

- Hoarseness

- Constipation

- Decreased sex drive

The most common cause of primary hypothyroidism in the United States is inflammation of the thyroid gland, an autoimmune condition called **Hashimoto's thyroiditis** or Hashimoto's disease. It occurs when the body's immune system mistakenly identifies cells in the thyroid as harmful substances and sends out white blood cells to attack and destroy the gland. Another less common type of primary hypothyroidism is autoimmune atrophic thyroiditis. In this condition, TSH-blocking antibodies halt the effect of TSH on the thyroid, causing the gland to waste away (**atrophy**). An enlarged thyroid (goiter) is a telltale sign of this endocrine disorder.

Hypothyroidism also can result from diets that are low in iodine. (See "The Goiter Belt.") The use of iodized salt has nearly eliminated this cause

The Goiter Belt

In the early 1900s, so many people in the Midwest and Great Lakes regions of the United States developed an enlarged thyroid gland that the two regions became known as the "goiter belt." The condition was widespread because people in those areas did not get enough iodine in their diets. Soil in the Midwest and Great Lakes regions was iodine-poor because thousands of years ago, glaciers stripped the essential element from the soil and washed it away.

To combat the problem, iodine was added to table salt beginning in the 1920s. Michigan was the first state to have access to iodized salt; by 1940, it was being used in households around the country. Although iodine deficiency is quite rare today due to the regular use of iodized salt, it does still exist, especially in underdeveloped parts of the world.

in modern countries. Other causes of hypothyroidism can include congenital defects, surgical removal of the thyroid gland, irradiation of the gland, or hyperthyroid medications that work too well.

If a physician suspects a thyroid problem, he or she will order blood tests to determine the levels of T_4 and serum TSH. In primary hypothyrodism, T_4 levels are low and TSH levels are high. The serum TSH is the easiest and most sensitive test for confirming the presence of primary hypothyroidism.

The symptoms of hypothyroidism are usually easily reversed with medication containing thyroid hormone. Most patients take the hormone for the remainder of their lives, although their dose may change from time to time.

HYPERTHYROIDISM

This condition is the opposite of hypothyroidism. Hyperthyroidism, or thyrotoxicosis, results when levels of thyroid hormones in the blood are excessively high. (*Note*: Some experts define thyrotoxicosis as a clinical condition that includes hyperthyroidism as a cause; however, this book follows the current standard of using the two terms interchangeably.) Excess thyroid hormones throw the body's metabolism into overdrive. Symptoms may include increased heart rate (**tachycardia**) and high blood pressure, weight loss, nervousness, tremor, excessive sweating, changes in the menstrual cycle, goiter, diarrhea, and heat intolerance. Patients with this condition are often described as being full of energy, or "hyperkinetic." Other patients may be able to eat whatever they want and never gain weight. Hyperthyroid children tend to outgrow their clothes rather quickly. Their permanent teeth grow in slower than usual. Other childhood symptoms can include delayed puberty, short-term memory loss, shaky hands, dry and itchy skin, and increased sensitivity to cold and heat.

The most common cause of hyperthyroidism is an autoimmune disorder called Graves' disease (diffuse toxic goiter). Unlike other autoimmune diseases that usually trigger *hypo*thyroidism, this condition occurs when the body's immune system sends out specific antibodies that thrust the thyroid gland into overdrive. Graves' disease patients have hyperthyroidism and one or more of the following symptoms: protruding eyes; goiter; a diffuse, nonpitting edema; and thickening of the skin on the lower legs and feet (**pretibial myxedema**). The disease most often strikes women between 20 and 40 years of age. Barbara Bush, former first lady of the United States, and

Portrait of Robert James Graves. © National Library of Medicine.

U.S. Olympic athlete Gail Devers are among the millions who seek treatment for this thyroid disorder. (See "The Famous Faces of Graves' Disease.") Late-stage Graves' patients often have bulging eyes and wide-eyed stares (a condition called exophthalmos) that are caused when the disease's instigating antibodies stimulate tissues other than the thyroid. Vision damage can be permanent.

The abrupt onset of dangerous hyperthyroid symptoms is referred to as "thyroid storm." Confusion, severe tachycardia, fever, and cardiovascular collapse characterize this urgent medical disorder. This life-threatening condition, which is almost always caused by Graves' disease, results from untreated or inadequately treated hyperthyroidism. It can be triggered by infection, trauma, a surgical procedure, embolism, or pregnancy complications. *Thyroid storm requires immediate treatment.*

In some cases, hyperthyroidism can result from improper doses of medication. For example, hypothyroid patients taking too much thyroid hormone can find themselves making a 180-degree turn into hyperthyroidism. Thyroid nodules can also trigger hyperthyroidism. (See entry in the latter part of this chapter.)

Blood tests can easily diagnose a hyperthyroid individual. In primary hyperthyroidism (when the gland itself produces excess hormone without influence by the pituitary gland), TSH levels are low, and T_3 and T_4 levels are high. Experts say a serum TSH is the best first test, because TSH is almost

The Famous Faces of Graves' Disease

Graves' disease does not discriminate based on wealth, social standing, or power, as one president and first lady discovered. On May 4, 1991, while jogging on the grounds of Camp David, President George Bush felt short of breath. His heart was also beating irregularly. After being taken to Bethesda Naval Hospital, the president told his doctors that he had been feeling tired for the last few weeks and had recently lost several pounds. A physical examination revealed that he had a swollen thyroid gland. Doctors diagnosed his condition as Graves' disease. President Bush was treated with radioactive iodine, which destroyed part of his overactive thyroid. Before long, he was able to make a full recovery.

By the time the president was diagnosed, the first lady had already been battling Graves' disease for two years. By the time Mrs. Bush was diagnosed and began radiation therapy in 1989, she had already developed the signature bulging eyes of the disease. The fact that two members of the same family, not to mention members of the presidential family, were struck by Graves' disease is extremely unusual. The odds of both spouses developing the condition within a two-year period is around one in 3 million (some people even hinted at a conspiracy, a suggestion at which the Bush administration scoffed).

Other famous names have also been associated with Graves' disease. Actor Marty Feldman's renowned bulgy-eyed stare was attributed to the condition. In 2000, Women's National Soccer Team captain Carla Overbeck learned that she had the disease. And Olympic runner Gail Devers was diagnosed with Graves' disease in 1991. She has since made a full recovery and has gone on to win four gold medals.

always suppressed in hyperthyroid patients. A type of test called a radioiodine uptake scan is performed to verify Graves' disease.

The goal of treatment is to slow the thyroid down. It is important to treat the disease immediately—extended high levels of thyroid hormone not only burn fat but also bone and muscle (putting patients at risk of osteoporosis). The first course of action is usually to prescribe antithyroid medicines, such as propylthiouracil (PTU) and methimazole, to prevent the gland from producing too much hormone. β-blockers, a type of medication commonly used to control blood pressure, are often recommended to control a rapid heartbeat. They do not, however, treat the *cause* of hyperthyroidism.

Iodine therapy (not to be confused with radioactive iodine treatments) is used for the emergency management of thyroid storm but is generally *not* used for routine treatment, because it can result in permanent thyroid damage.

If oral drug therapy is unsuccessful, the physician may suggest destroying parts of the thyroid gland with radioactive iodine treatments. ^{131}I is the most common form of therapy for American hyperthyroidism patients, and it is the treatment of choice for Graves' disease. When radioactive iodine is administered, the thyroid gland (which is unable to tell whether or not the iodine is radioactive) gathers it just as it would gather nonradioactive iodine. Radioiodine builds up in the thyroid hormone–producing cells and stays there long enough to radiate the gland and slow thyroid production. Any unused radioiodine is excreted in the urine. Experts say there is no proof that radioiodine increases the risk of cancer or leads to higher birth defect rates among women who become pregnant later in life.

In extreme cases, surgery may be required to remove the malfunctioning thyroid gland. Postoperative recurrences range between 2 and 9 percent. Risks include permanent damage to the gland, which results in permanent hypothyroidiom.

Graves' patients with exophthalmos undergo additional therapies (including corticosteroids, orbital radiation, and orbital decompression), because elevated antibodies and not thyroid hormone levels cause the eye protrusions.

THYROID NODULES AND TUMORS

A nodule is a lump in the thyroid gland. A patient may notice a lump in his or her neck, or a doctor will detect it during a routine exam. Discovery of a thyroid nodule should not elicit alarm; according to the American Cancer Society, nearly all thyroid nodules are benign.

If a nodule is present, a physician may first order an ultrasound to examine the lump. Depending on the outcome, either fine needle aspiration

Thyroid Cancer Symptoms

Early thyroid cancer often does not cause symptoms. But as the cancer grows, symptoms may include:

- A nodule in the front of the neck near the Adam's apple

- Hoarseness or difficulty speaking normally

- Swollen lymph nodes, especially in the neck

- Difficulty swallowing or breathing

- Pain in the throat or neck

It is important to note that these symptoms are not definitive signs of thyroid cancer and may be caused by a number of other conditions. Only a doctor can diagnose and treat the problem.

Source: National Cancer Institute, 2002.

or a biopsy may be performed to determine whether the lump is malignant or benign. If no cancer cells are found, the doctor may suggest medication to shrink the nodule or surgical removal of the lump itself.

In the rare instance that a nodule is cancerous (less than 5 percent are), additional treatment will be recommended depending on the tumor type. Thyroid cancer can occur at any age. Patients who received radiation therapy as a child or who have undergone radiation therapy to the neck are at increased risk. (Radiation therapy was commonly used in the 1950s to treat enlarged thymus glands, adenoids and tonsils, and skin disorders.) Other risk factors include a family history of thyroid cancer and a chronic enlargement of the thyroid gland (see "Thyroid Cancer Symptoms").

There are four general types of thyroid cancer: papillary, follicular, medullary (solid, with amyloid struma), and undifferentiated and anaplastic (rare).

Papillary Carcinoma

Also called papillary cancer or papillary adenocarcinoma, this is the most common thyroid cancer; it accounts for more than half of all thyroid cancers. This tumor type strikes women two to three times more often than men. Papillary cancers are more common in young patients, especially those with a history of radiation exposure. The tumors develop from the thyroid follicle cells and progress quite slowly. More often than not, they occur in only one lobe of the thyroid gland. The cause of this cancer is unknown; however, mutations that affect cell growth in the thyroid may play a role. Research suggests these cancers may be TSH-dependent and might develop in goiters secondary to Hashimoto's thyroiditis.

Despite its slow-growing status, papillary cancer usually spreads early to the lymphatic system in the neck. Despite this metastasis, 95 percent of adults with this cancer survive ten years (NLM, 2002). The prognosis is even better in patients under age 40 and for those who have tiny tumors.

Follicular Carcinoma

According to the National Library of Medicine, follicular cancer is the second most common type of thyroid cancer, accounting for about 30 percent of all cases. This type of cancer occasionally spreads to the lungs, brain, liver, bladder, skin, and bone, but, unlike papillary carcinomas, follicular carcinomas rarely spread to the lymph nodes. Follicular cancer occurs in a slightly older age group than papillary carcinoma; it is also less common in children. A patient's age has a great bearing on his or her prognosis. Those over 40 tend to have a more aggressive form of cancer; tumors in these patients usually do not respond as well to iodine treatment as tumors found in younger patients.

Medullary Thyroid Carcinoma (MTC)

MTC is a rare type of thyroid cancer. According to the National Cancer Institute, it accounts for 5 to 10 percent of thyroid cancer cases. Unlike papillary and follicular carcinomas, which arise from thyroid hormone–producing cells, MTC originates from nonthyroid cells called C cells. These cancers produce the hormone calcitonin (see Chapter 4) and a protein called **carcinoembryonic antigen (CEA),** which is also found in other cancers.

MTC has a much lower cure rate than common thyroid cancers; however, overall ten-year survival rates are 90 percent when the cancer is confined to the thyroid gland. That rate drops to just 20 percent when metastasis to distant sites occurs.

MTC presents itself in four different clinical ways:

Sporadic: The first and most common type of medullary cancer is called sporadic MTC. This cancer accounts for 80 percent of all medullary thyroid cancers. Tumors are typically unilateral, meaning they occur most commonly in just one thyroid lobe. Sporadic MTC is most common in people ages 40 to 60 years, with women patients outnumbering men by a 3:2 ratio. Diarrhea is a common symptom because the hormones secreted by the tumors cause increased gastrointestinal secretions. Sporadic MTC is not inherited.

MEN 1 (Wermer's Syndrome): When medullary thyroid cancer runs in a family, it is referred to as "familial medullary thyroid cancer" or "multiple endocrine neoplasia (MEN) syndrome." MEN syndromes are a group of inherited endocrine disorders that occur together in one patient, although not necessarily at the same time. MEN 1 is a rare, inherited disorder that strikes men and women equally, most often between the ages of 40 and 50. Doctors sometimes refer to this syndrome as multiple endocrine adenomatosis or Wermer's syndrome, after American internist Paul Wermer (1898–1975), one of the first doctors to recognize it. Tumors strike the parathyroid gland first and most often, followed by the pituitary gland and pancreatic islet cells.

MEN 2a (Sipple Syndrome): MEN 2a (Sipple Syndrome) is a hereditary disorder in which tumors develop in two or more of the following endocrine glands: thyroid, adrenal, and/or parathyroid. It is caused by a defective gene called RET, which helps control the normal growth of endocrine tissues. (If medullary thyroid cancer is suspected, a doctor can order blood tests to check for the presence of this altered gene.) Males and females are equally affected. The disorder may occur at any age, although it is more common in patients in their 30s.

MEN 2b: This syndrome can be, but is not always, inherited. Patients with this syndrome usually develop thyroid and adrenal tumors (**phenochromocytoma**), but the most distinctive feature, in most patients, is the presence of benign nerve tissue growths on the tongue and lips. The eyes also are commonly involved. Stomach problems, such as constipation, diarrhea, and, occasionally, **megacolon,** are common; theoretically they result from intestinal **ganglioneuromatosis.** MEN 2b patients usually develop medullary carcinoma in their 30s; men and women are equally affected. MTC is most aggressive in the MEN 2b syndrome.

Anaplastic Carcinoma

Anaplastic carcinoma is a rare but very aggressive and invasive form of thyroid cancer. Doctors believe it develops from an existing papillary or follicular cancer. Its rapid spread throughout the neck and the rest of the body makes it extremely fatal. Patients with the giant cell variety of this cancer have an expected life span of less than six months; those with the small cell variety may live up to five years after diagnosis. Anaplastic carcinoma usually occurs in people over 60 years and may cause obstruction of the trachea. Persistent hoarseness, changing voice, cough, or coughing up blood may indicate the presence of this disease. The National Library of Medicine estimates that anaplastic carcinoma accounts for just 1 percent of all thyroid cancers, striking three out of every 10,000 people.

Thyroid Lymphoma

Lymphomas of the thyroid gland are very uncommon. They occur from lymphocytes, the main cell type of the immune system. Unlike most lymphomas, which begin in the lymphocytes, **extranodal lymphomas** grow outside the lymph nodes in organs such as the thyroid. Most thyroid lymphomas occur in people with Hashimoto's thyroiditis (see explanation earlier in this chapter).

If the diagnosis is thyroid cancer, the doctor must determine the stage, or extent, of the disease to plan the most successful therapy. Ultrasonography, MRI, or CT are useful tools for visualizing cancer spread.

Thyroid cancer treatment involves a three-part course of action—surgery, radioactive iodine, and medication. During surgery, endocrinologists will remove as much of the tumor as possible; tumor size dictates how much of the gland is removed. Frequently, the entire thyroid gland is removed. (If surgery is not possible, external radiation therapy may be performed.) Chemotherapy, with or without radiation therapy, may be indicated for cancers that have metastasized. Postsurgery, most patients are treated with radioactive iodine and then with a thyroid hormone medication called levothyroxine sodium. They will remain on this medication for life. Routine follow-up typically involves blood tests every few months and a ^{131}I scan at nine to twelve months and then yearly.

HYPERPARATHYROIDISM

Every year, about 100,000 Americans develop primary hyperparathyroidism, a condition that occurs when one or more of the parathyroid glands (see Chapter 2) go into overdrive and produce too much parathyroid hormone (PTH). The imbalance triggers an excessive transfer of calcium from the bones to the blood, giving way to a condition called **hypercalcemia.**

Bones lose calcium, putting patients at risk of osteoporosis. Calcium levels in the urine increase, causing kidney stones. In most cases, hyperparathyroidism is caused by a benign tumor on one of the parathyroid glands. But sometimes, excess PTH production is due to two or more enlarged parathyroid glands, a condition called hyperplasia. In very rare instances, hyperparathyroidism is caused by cancer of a parathyroid gland. The disease is more prevalent in women; females with hyperparathyroidism outnumber males 2 to 1. Risk also increases with age. Two out of 1,000 women aged 60 years and older will develop hyperparathyroidism. Few cases are inherited. However, MEN1 is one rare inherited syndrome affecting the parathyroid glands.

Symptoms of hyperparathyroid patients may be severe, subtle, or nonexistent. Today, sensitive blood tests that screen for a wide range of conditions, including high calcium levels, can identify asymptomatic patients. Most symptoms, when they do appear, are mild and nonspecific. Weakness, fatigue, depression, and aches and pains can mimic a myriad of other diseases, a quandary that often hampers diagnosis. Patients with more advanced cases of the disease may present with appetite loss, nausea, vomiting, constipation, confusion or impaired thinking and memory, increased thirst and urination, and osteoporosis. Patients are at increased risk of developing peptic ulcers, high blood pressure, and pancreatitis.

Hyperparathyroidism is easily diagnosed with a blood test that measures parathyroid hormone levels. Doctors may also check for high blood calcium levels. If a blood test suggests hyperparathyroidism, imaging tests may be performed to measure bone density (to assess bone loss) and to scan for the presence of kidney stones. A twenty-four-hour urine collection may also be ordered so the physician can assess the risk of kidney damage.

Surgical removal of the enlarged gland (or glands) is the only treatment for hyperparathyroidism. It is successful in 95 percent of cases. However, an advisory board convened by the National Institutes of Health in 1990 suggested that some patients with mild cases of the disease might not require immediate treatment. Asymptomatic patients with slightly elevated blood calcium levels and no kidney or bone disease might consider watchful waiting. The panel recommended monitoring the patient's calcium levels and kidney function every six months, performing an annual abdominal x-ray, and measuring bone mass after one to two years.

Parathyroid autotransplantation may be an option for patients in whom all four parathyroid glands are affected or who require several surgeries. Parathyroid autotransplantation involves the removal of all parathyroid tissue in the neck. A small amount of the tissue is then transplanted into the patient's forearm, where it remains and continues to produce parathyroid hormone for the body.

HYPOPARATHYROIDISM

Hypoparathyroidism is marked by a deficiency of parathyroid hormone (PTH). The disorder is usually caused by injury to the parathyroid glands during head and neck surgery. It is rarely a side effect of radioactive iodine treatment for hyperthyroidism. PTH secretion may also be disrupted when blood levels of magnesium are low or when blood pH is too high, a condition called metabolic alkalosis. The incidence is quite low; only four in 100,000 people develop hypoparathyroidism.

The lack of PTH causes blood calcium levels to fall and phosphorus levels to rise. Low blood calcium levels may cause symptoms such as tingling of the lips, fingers, and toes and muscle cramps or spasms. Hypoparathyroid patients might also complain of abdominal pain, dry hair, brittle nails, and dry skin. They may develop cataracts, and children might have weakened tooth enamel. Female patients sometimes experience painful menstrual cycles. Convulsions or seizures can occur in severe cases.

When hypoparathyroidism is the result of a congenital defect, it is called DiGeorge syndrome. Familial hypoparathyroidism occurs with other endocrine diseases such as adrenal insufficiency in a syndrome called type I polyglandular autoimmune syndrome.

If hypoparathyroidism is suspected, the doctor will order blood tests to measure calcium, phosphorous, and parathyroid hormone levels. Treating this disease involves restoring the calcium and associated mineral balance within the body. Patients will usually take oral calcium carbonate and vitamin D supplements for their entire lives. Blood levels will be checked periodically to ensure proper dosage. Endocrinologists recommend a high-calcium, low-phosphorous diet.

When diagnosed early, hypoparathyroidism prognosis is good. However, dental changes and cataracts are irreversible.

Diseases of the Adrenal Gland

Steroid hormones produced by the adrenal gland affect nearly every system of the body to some degree. Disturbances in the production of these hormones can lead to a myriad of diseases, with symptoms ranging from physical stress to abnormal cholesterol levels. Like every other endocrine gland, the adrenal gland has deficiency-related syndromes and hormonal-excess disorders.

ADRENOCORTICOL INSUFFICIENCY

Adrenocorticol insufficiency (AI) is a condition marked by a decreased function of the adrenal cortex and, as a result, insufficient production of adrenal corticosteroid hormones. Failure to produce adequate levels of cortisol, or adrenal insufficiency, can occur for different reasons. Primary AI results when at least 90 percent of the adrenal cortex has been damaged. Secondary AI results when the cortex has wasted away due to insufficient ACTH. Primary AI is most commonly referred to as Addison's disease, named after Thomas Addison, the man who identified the disorder in 1849 (see Chapter 8). The rare endocrine disease affects about 1 in 100,000 people, regardless of age or gender. President John F. Kennedy had Addison's disease (see photo). Addison's disease occurs when levels of cortisol (a stress hormone) and, in some cases, the hormone aldosterone (which regulates blood pressure and the body's water/salt balance) are too low. The disease is sometimes called hypocortisolism.

Symptoms usually begin gradually, mimicking an array of other conditions. However, they can appear suddenly. Patients with Addison's disease

John F. Kennedy as he delivers his Address on Civil Rights on radio and television, June 11, 1963. © Abbie Rowe, National Park Service/John Fitzgerald Library, Boston.

tend to be tired all the time; muscle weakness and lethargy are extremely common. Weight loss and decreased appetite are typical, and digestive disturbances such as nausea, vomiting, and diarrhea occur in almost 50 percent of cases. The lack of aldosterone leads to low blood pressure; levels fall even further when the patient is standing, causing dizziness or even fainting. Addison's patients tend to have a dark tanned appearance (hyperpigmentation), especially on scars, skin folds, pressure points such as the elbows, knees, knuckles, and toes, the lips, and mucous membranes. The disease can make a person feel irritable and depressed. Because low levels of aldosterone cause salt loss, a patient may crave salty foods. Children with this disease tend to have severe hypoglycemia. Women may notice cessation of their menstrual periods.

Often, symptoms are ignored until a stressful event causes them to worsen. This condition, called Addisonian crisis, is precipitated by severe diarrhea and vomiting and coincides with shock and loss of consciousness. Some patients may experience a sudden penetrating pain in their lower back, abdomen, or legs. In about a quarter of patients, symptoms are first noted during an Addisonian crisis. Left untreated, this severe form of Addison's disease can be fatal.

In its early stages, an adrenal insufficiency disorder can be difficult to diagnose. If a physician suspects Addison's disease, he or she will order blood and urine samples. Laboratory tests will be done to determine whether there are insufficient levels of cortisol and then to establish the cause of the deficiency. The "short" ACTH test is the most specific method for diagnosing Addison's disease. In this test, blood and/or urine cortisol levels are measured before and after a synthetic form of ACTH is given by injection. Healthy patients will exhibit a rise in blood and urine cortisol levels. Those with either form of adrenal insufficiency will respond poorly or will not respond at all.

When the short ACTH test results are abnormal, a "long" ACTH stimulation test is performed to reveal the cause of adrenal insufficiency. In this test, synthetic ACTH is injected either intravenously or intramuscularly over a two- to three-day period. Blood and/or urine cortisol levels are measured the day prior to the test and during the injection period. Patients with primary adrenal insufficiency will not produce cortisol during this time; however, patients with secondary adrenal insufficiency will have adequate responses to the test on the second or third day.

In patients suspected of having an Addisonian crisis, injections of salt solutions and steroid hormones begin immediately. Diagnosis is made later, because the substances used in treatment interfere with diagnostic blood tests. Once the crisis is controlled and medication has been stopped, the doctor will delay testing for up to one month to obtain an accurate diagnosis.

Once a diagnosis of primary adrenal insufficiency has been made, x-ray exams of the abdomen and a CT scan may be taken to determine its cause. X-rays can spot calcium deposits on the adrenals; calcium deposits may indicate tuberculosis (TB). TB accounts for about 20 percent of cases of primary adrenal insufficiency in developed countries. CT scans reveal the size and shape of the pituitary gland.

Treatment of Addison's disease involves replacing the hormones that are not being produced by the adrenal glands. Hormone replacement therapy will control the symptoms of the disease, but the treatment must continue throughout a patient's life. Patients are usually instructed to take a dose in the morning and another in the late afternoon to mimic the body's normal rhythm of steroid production. The body's natural cortisol is replaced with oral hydrocortisone tablets (a synthetic glucocorticoid). If aldosterone is also deficient, it is replaced by a daily dose of a mineralocorticoid called fludrocortisone acetate. Because patients with secondary AI normally maintain aldosterone production, aldosterone replacement therapy is usually unnecessary.

OVERACTIVE: CUSHING'S SYNDROME

When the adrenal glands release an excess of glucocorticoids, Cushing's syndrome, or hypercortisolism, can develop. In children, Cushing's syndrome most often results from large doses of synthetic corticosteroid drugs (such as prednisone) used to treat ailments like asthma or lupus. Pituitary adenomas (see Chapter 11) are the most common cause of Cushing's syndrome. These noncancerous tumors give off increased amounts of ACTH. Most patients have just one tumor. This form of the syndrome, called Cushing's disease, affects women five times more frequently than men.

Sometimes benign or malignant tumors that develop outside the pituitary can produce ACTH and trigger Cushing's syndrome. The most common forms of ACTH-producing tumors are small-cell lung cancer and carcinoid tumors.

Pancreatic islet cell tumors (see Chapter 13) and medullary carcinomas of the thyroid (see Chapter 10) can also produce ACTH. A noncancerous adrenal tumor can also cause Cushing's syndrome. Adrenocortical carcinomas, or adrenal cancers, are the least common cause of Cushing's syndrome. These tumors usually lead to very high hormone levels and rapid development of symptoms.

Most cases of Cushing's syndrome are not inherited. Rarely, however, some people inherit a tendency to develop tumors of one or more endocrine glands. This is called familial Cushing's syndrome.

Cushing's syndrome is exceptionally rare; an estimated 10 to 15 out of every million people are affected each year. It most commonly affects adults aged 20 to 50. Symptoms may take several years to develop. Most patients display a rounded face, overweight torso, increased fat around the neck, and thinning of the arms and legs. Children tend to be obese and grow slowly. Patients might notice they bruise quite easy. Muscle weakness, acne, high blood pressure, hyperglycemia, and psychological changes (such as anxiety and depression) are common. Bones are quite weak and even routine activities like rising from a chair can cause rib and spinal fractures. A woman may display excess hair growth on her face, neck, chest, abdomen, and thighs. Menstrual periods may become irregular or stop. Men with the condition are usually impotent.

Diagnosis of Cushing's syndrome involves a battery of laboratory tests. The twenty-four-hour urinary free cortisol test is the most specific diagnostic test. The patient's urine is collected (usually at home) over a twenty-four-hour period and tested for cortisol. Cortisol levels over 50–100 micrograms a day (in an adult) suggest Cushing's syndrome.

Additional tests are performed to reveal the exact location of the cortisol-releasing abnormality. The dexamethasone suppression test helps distinguish patients with excess production of ACTH due to pituitary adenomas from those with ectopic ACTH-producing tumors. In this test, patients take varying doses of dexamethasone, a synthetic glucocorticoid, orally every six hours for four days. Then another twenty-four-hour urine test is performed and the cortisol level is measured. Because cortisol and other glucocorticoids signal the pituitary to lower its secretion of ACTH, the normal response after taking dexamethasone is a drop in blood and urine cortisol levels. Responses will differ depending on whether the cause of Cushing's syndrome is a pituitary adenoma or an ectopic ACTH-producing tumor. The test is not foolproof, however. Some drugs can cause false-negative results. False-positive results are common in patients with depression, alcohol abuse, high estrogen levels, acute illness, and stress.

The CRH stimulation test is used to differentiate between pituitary adenomas and ectopic ACTH syndrome or cortisol-secreting adrenal tumors. Patients are given an injection of CRH, the corticotropin-releasing hormone that causes the pituitary to secrete ACTH. Patients with pituitary adenomas usually display a rise in blood levels of ACTH and cortisol. This response is rarely seen in patients with ectopic ACTH syndrome and patients with cortisol-secreting adrenal tumors.

Petrosal sinus sampling helps separate pituitary from ectopic causes of Cushing's syndrome by determining whether the hormone is originating from the pituitary gland. Blood samples are drawn from the petrosal sinuses, the veins that drain the pituitary. The procedure is performed under local anesthesia and mild sedation. Levels of ACTH are measured and compared with ACTH levels in a forearm vein. ACTH levels higher in the petrosal sinuses than in the forearm vein indicate the presence of a pituitary adenoma; similar levels suggest ectopic ACTH syndrome.

Depending on the specific cause, doctors may treat the condition with surgery, radiation therapy, chemotherapy, or hormone production–blocking drugs. When symptoms result from a large dose of steroid hormone medication, doctors usually gradually decrease the dosage. If the culprit is an ACTH-secreting pituitary adenoma, surgical removal of the tumor, known as transsphenoidal adenomectomy, is recommended. According to the National Institutes of Health, 80 percent of patients are cured when an experienced surgeon performs the surgery. If results are unfavorable, surgery can be repeated, often with good results. After curative pituitary surgery, patients display a temporary drop in the production of ACTH. To compensate, patients are given a synthetic form of cortisol. Most patients can stop this replacement therapy within a year.

For patients in whom surgery has failed or has not been recommended, radiotherapy is an alternative treatment. Radiation to the pituitary gland is given over a six-week period. Almost half of adults and up to 80 percent of children see improvement after radiation therapy.

To cure excess cortisol levels caused by ectopic ACTH syndrome, doctors must eliminate all of the cancerous, ACTH-secreting tissue. The choice of cancer treatment depends on the type of cancer and how far it has spread. Surgery is the mainstay of treatment for benign as well as cancerous adrenal tumors. In those with familiar Cushing's syndrome, surgical removal of the adrenal glands is required.

TUMORS OF THE ADRENAL GLAND

This chapter has already discussed the adrenal adenoma that leads to Cushing's disease. There are three other types of adrenal tumors. The first one, an extremely rare tumor, produces male or female sex hormones and can trigger feminine qualities in men and masculization in women. A second type, called pheochromocytoma, blossoms in the core of the adrenal gland and secretes too much epinephrine (adrenaline) and norepinephrine (noradrenaline). The last type of adrenal tumor occurs in the adrenal cortex and secretes too much aldosterone. This condition is called Conn's syndrome. In all cases, symptoms consist of high blood pressure, excessive sweating, increased heart rate, weight loss, constipation, and personality changes. Pheochromocytomas typically run in families and may be associated with thyroid cancer. Too much sodium and too little potassium in the blood are hallmarks of Conn's syndrome. If left untreated, pheochromocytoma can be life-threatening. In all cases, surgery to remove the adrenal gland is recommended.

CONGENITAL ADRENAL HYPERPLASIA

Congenital adrenal hyperplasia (CAH) is caused by a genetic abnormality that blocks the manufacture of steroid hormones by the adrenal glands. Patients with CAH lack the enzyme the adrenal glands need to produce cortisol and aldosterone. Because synthesis of these hormones is blocked, the adrenal glands begin overproducing their other hormone product—androgens (male sex hormones). The disease most often strikes infants and children; it is quite rare in adults. About one out of every 10,000 to 18,000 children are born with this disease.

A newborn girl will exhibit a swollen clitoris with ambiguous genitalia (although internal female structures are normal). She may appear more male than female. Masculine features appear as the female grows older. Symptoms may include deepening of the voice, the appearance of facial hair, and failure to menstruate at puberty.

In baby boys, no obvious abnormality is present at birth. But an affected male will appear to be entering puberty when he is just a toddler. The penis enlarges, pubic hair starts to grow, the voice deepens, and muscle mass increases.

Sometimes, a baby is born with an extremely severe form of congenital adrenal hyperplasia, triggering an adrenal crisis. The baby will be very sick. Symptoms of adrenal crisis include vomiting, dehydration, electrolyte changes, and an irregular heartbeat. Untreated, this condition can lead to death within one to six weeks after birth.

The physician will perform a battery of tests to diagnose the condition and rule out other disorders. Laboratory tests will hunt for decreased levels of aldosterone, cortisol, and other steroidal hormones in the blood, as well as for abnormal salt levels in both the blood and urine. An x-ray may be taken to reveal bone age. Patients with this disorder have a markedly advanced bone age. Daily doses of cortisol (dexmethasone, fludrocortisone, or hydrocortisone) can help return hormone levels to normal. A baby girl with masculine genitalia may require reconstructive surgery during infancy.

Diseases of the Endocrine Pancreas

While many think of the pancreas as an organ, it is, in fact, both an exocrine and an endocrine gland. The islets of Langerhans (see Chapter 5) comprise the endocrine portion of the pancreas, although in essence they total only a small fraction of it's weight. The thousands of tiny islets produce a number of substances (including insulin) that are important to endocrine function. As with all endocrine organs, too little or too much of these substances (hormones) can wreak havoc on the body.

DIABETES MELLITUS

Diabetes mellitus is a serious, lifelong endocrine disorder characterized by a deficiency of insulin and an excess of glucagon. Most ingested food is broken down into glucose, the main source of fuel for the body. After digestion, glucose passes into the bloodstream and is sent to cells, which need it to grow. The pancreas normally produces a level of insulin necessary to trigger this mechanism. In people with diabetes, however, the pancreas either produces little or no insulin, or the cells do not respond appropriately to the insulin that is produced. As a result, blood sugar levels rise and spill into the urine, passing out of the body. Diabetes is widely recognized as one of the leading causes of death and disability in the United States. The American Diabetes Association estimates that about 17 million Americans have diabetes; a third of them are unaware that they are sick. There are three major types of diabetes, each one with different symptoms and a different cause. They are:

1. Type 1 diabetes

2. Type 2 diabetes

3. Gestational diabetes

Type 1 Diabetes

This type of diabetes (also referred to as insulin-dependent diabetes mellitus or IDDM) used to be called juvenile diabetes, because onset typically occurs in childhood or, sometimes, young adulthood. Although most type 1 diabetes cases appear between ages 10 and 14, the disease can strike at any time. Type 1 diabetes accounts for 5 to 10 percent of diagnosed diabetes cases in the United States. Type 1 diabetes is an autoimmune disease, meaning that the immune system attacks a part of the body, in this case the insulin-producing beta cells of the pancreas. Patients with this type of diabetes do not produce insulin. It is unknown why this autoimmune diabetes develops. Sometimes it follows a viral infection such as the flu, mumps, rubella, measles, encephalitis, polio, or Epstein-Barr virus. Some people are more genetically prone to developing the disease this way. In rare cases, injury to the pancreas can trigger type 1 diabetes.

Symptoms of diabetes mellitus are typically the same, regardless of the type. They include increased thirst (polydipsia) due to dehydration, increased urination (polyuria) due to high levels of glucose (a diuretic) in the urine, and weight loss despite increased appetite (polyphagia). Diabetics feel constantly hungry because the areas of the hypothalamus that control appetite have insulin-sensitive transport systems. Other symptoms include nausea, vomiting, vaginitis, skin infections, frequent bladder infections, visual disturbances, and constant fatigue. The disease can cause multiple long-term complications including kidney problems, pain from nerve damage, blindness, and early coronary heart disease and stroke.

If not diagnosed and treated with insulin, a person can lapse into a life-threatening diabetic coma, also known as diabetic ketoacidosis. Some researchers believe this acute complication can also occur when a diabetic is under stress from illness or injury. A person in a diabetic coma will experience increased urination and an unquenchable thirst. The patient may appear weak or drowsy and exhibit a flushed complexion. He or she may vomit, have diarrhea, or experience abdominal pain. Sometimes the breath takes on a sweet smell that can mimic the odor of alcohol (this is actually acetone being expelled by the lungs). As the condition progresses, breathing becomes faster and deeper. According to the American Diabetes Association, ketoacidosis causes about one in ten diabetic deaths in people with diabetes who are under age 45.

Type 2 Diabetes

Formerly called adult-onset diabetes or noninsulin-dependent diabetes mellitus (NIDDM), type 2 diabetes is the most common form of the disease. About 90 to 95 percent of diabetics have this type. It typically develops in adults aged 40 and older and is most common in adults over age 55. Ethnicity greatly affects the risk of developing the disease. Risk factors are described in "Who Is at Risk for Type 2 Diabetes?"

Type 2 diabetes usually begins with insulin resistance, a condition in which fat, muscle, and liver cells do not use insulin properly. In the beginning, the pancreas can keep up with the added demand by producing more insulin. But over time, it loses the ability to secrete enough insulin in response to rising blood glucose levels. After several years, insulin production decreases. Being overweight and inactive increases one's odds of developing type 2 diabetes. Although a symptom of type 2 diabetes is weight loss, about 80 percent of patients are overweight, including children. Endocrinologists theorize that excess body fat may play a role in insulin resis-

Who Is at Risk for Type 2 Diabetes?

You are more likely to develop type 2 diabetes if:

- You are overweight. Type 2 diabetes is more common in people who are overweight.

- You are 45 years old or older.

- You have an immediate family member with diabetes.

- Your family background is African American, American Indian, Asian American, Hispanic American/Latino, or Pacific Islander.

- You have had gestational diabetes or have given birth to a large baby (over nine pounds).

- You have high blood pressure.

- Your HDL ("good") cholesterol is 35 or lower, or your triglyceride level is 250 or higher.

- You exercise fewer than three times a week.

Source: National Diabetes Information Clearinghouse, 2003.

tance. Type 2 diabetes is part of a metabolic syndrome that includes obesity, hypertension, and increased levels of blood lipids.

The other symptoms (and possible complications) of type 2 diabetes are basically the same as those of type 1. Unlike type 1 diabetics who experience a sudden onset of symptoms, type 2 diabetics develop symptoms gradually. Unfortunately, the skyrocketing rates of obesity among children and adolescents over the last three decades, coupled with increasingly sedentary lifestyles, have contributed to a sharp rise in type 2 diabetes cases in young people. U.S. government projections indicate that there will be a 165 percent increase in diabetes by the year 2050.

Gestational Diabetes

Gestational diabetes is a form of diabetes present only in pregnant women who have never displayed increased blood sugar levels. About 135,000

What Is Your Risk for Gestational Diabetes?

		Yes	No
1.	Are you Hispanic, African American, Native American, South or East Asian, Pacific Islander, or Indigenous Australian?	❏	❏
2.	Are you overweight or obese?	❏	❏
3.	Does anyone in your family have diabetes now, or have they had diabetes in their lifetime?	❏	❏
4.	Are you older than 25?	❏	❏
5.	Did you have gestational diabetes with a past pregnancy?	❏	❏
6.	Have you had a stillbirth or a very large baby with a past pregnancy?	❏	❏

- Answering YES to TWO or more questions suggests a high risk. You should be tested as soon as you know you are pregnant.

- Answering YES to ONLY ONE question suggests an average risk. You should be tested when you are between twenty-four and twenty-eight weeks pregnant.

- Answering NO to ALL questions suggests a low risk. Testing is unnecessary unless your doctor advises otherwise.

Source: National Institute of Child Health and Human Development, 2000.

pregnant women develop the condition every year, making it one of the top health concerns related to pregnancy (see "What Is Your Risk for Gestational Diabetes?"). Gestational diabetes usually appears in the fifth or sixth month of pregnancy. Most of the time, gestational diabetes disappears and the body returns to normal after childbirth. In women with gestational diabetes, glucose can't move into the body's cells. The result is a high level of sugar in the bloodstream. Symptoms mimic those of the other forms of diabetes.

Left untreated, gestational diabetes can cause pregnancy complications and birth defects. Children whose mothers had gestational diabetes are at increased risk for Respiratory Distress Syndrome (RDS), obesity, and adult-onset diabetes.

Diabetes Complications

It is incredibly important to keep diabetes in check. Untreated, the disease increases the risk of long-term complications that affect almost every part of the body. According to the National Institutes for Health, diabetes often results in serious problems of the eyes, heart, kidneys, and nerves. When blood glucose levels are high, a refractive error in the eye can develop, causing blurriness and sometimes blindness. Rapid lowering of blood sugar levels can make blurriness worse; glasses should not be fitted until blood sugar has been stable for almost two months. Kidney disease occurs from long-term deterioration of the small blood vessels. Diabetic neuropathy is a deterioration of nerve fiber function. Often, diabetics note a tingling sensation in their feet and hands. This is due to nerve fiber damage. According to the American Diabetes Association, one in five diabetics visits the hospital for foot problems. Bad feet are a major cause of disability in patients with diabetes. Diabetes-related foot problems usually result from poor circulation, nerve damage, and decreased resistance to infection. Diabetes can also cause changes in the bone structure and soft tissue of the feet. Undoubtedly, the most serious problem caused by diabetes is heart disease. Diabetics are more than twice as likely as those without diabetes to have heart disease or a stroke.

Diabetes represents a substantial cost burden to society. A 2003 study by the American Diabetes Association found that the annual cost of diabetes climbed from $98 billion in 1997 to $132 billion in 2002. Indirect costs resulting from lost workdays, restricted activity days, mortality, and permanent disabilities related to diabetes totaled nearly $40 billion yearly. According to the study, the U.S. government spends $13,243 on each diabetic patient, compared to $2,560 per person for those who don't have diabetes.

Testing and Diagnosis

The American Diabetes Association recommends that everyone over 45 years of age be tested regularly for diabetes. Earlier testing is advised if you have any of the following risk factors:

- A weight that is 20 percent more than ideal body weight
- High blood pressure
- Low HDL cholesterol levels (under 35 mg/dl) and high triglyceride levels (over 250 mg/dl)
- A close relative with diabetes
- A high-risk ethnic group background
- Delivered a baby weighing over nine pounds
- A history of gestational diabetes

Blood and urine tests can be performed to measure the concentration of sugar. Glucose in the urine is called "glycosuria." A high amount of sugar in the bloodstream is termed "hyperglycemia." These findings are evidence of diabetes.

The fasting plasma glucose test is the preferred test for diagnosing type 1 or type 2 diabetes. A diagnosis of diabetes is made if the fasting plasma glucose value is 126 mg/dl or more. The oral glucose tolerance test (OGTT) is the most common method of diagnosis for gestational diabetes, although it can also be used to diagnose the other forms of the disease. If the doctor suspects diabetes, the patient will be instructed to consume a drink containing glucose dissolved in water. Plasma glucose levels are measured over a three-hour period. In nonpregnant patients, a plasma glucose value of 200 mg/dL or more in the blood sample taken two hours after a person has consumed the drink indicates diabetes. The threshold values for pregnant women are lower, because glucose levels normally drop during pregnancy.

Prediabetes

People with glucose levels between "normal" and "diabetes" are called prediabetics. About 16 million people, ages 40 to 74, are considered prediabetic; they are at increased risk of developing diabetes, heart disease, and stroke. In July 2003, researchers at Johns Hopkins University in Baltimore, Maryland, reported that prediabetics may also be at increased risk of dying from cancer. There are two forms of prediabetes: impaired fasting glucose (IFG) is indicated when the fasting plasma glucose level is higher than normal but less than the level indicating a diagnosis of diabetes. Impaired glucose tolerance (IGT) means that the blood glucose level taken during the oral glucose tolerance test exceeds the normal limit but is not high enough for a diagnosis of diabetes.

Treating Diabetes

The goal of diabetes management is to keep blood glucose levels as close to the normal range as possible. The American Diabetes Association recommends that premeal blood sugar levels register 80 to 120 mg/dl and bedtime blood levels fall in the range of 100 to 140 mg/dl (see photo in color insert). Type 1 diabetics must take insulin daily to live. They can take insulin shots or use an insulin pump. To date, there are six main types of insulin. Scientists are working on developing patches that deliver insulin directly through the skin into the bloodstream. In July 2003, the U.S. Food and Drug Administration (FDA) cleared the first device for diabetics that integrates a glucose meter and an insulin pump with a dose calculator. The new device is expected to pave the way for a fully automated glucose monitoring and insulin delivery system.

There are numerous types of pills to treat type 2 diabetes. Each relies on a different mechanism to lower blood glucose. Combination pills put together two or more types of medicine. The types of drugs available include:

- Sulfonylureas stimulate the pancreas to make more insulin.
- Biguanides reduce the amount of glucose made by the liver.
- Alpha-glucosidase inhibitors slow the absorption of starches (which convert to sugars).
- Thiazolidinediones make a patient more sensitive to insulin.
- Meglitinides stimulate the pancreas to produce more insulin.
- D-phenylalanine derivatives prompt the pancreas to speed up insulin production.

In addition to prescription medications, treatment for diabetes mellitus includes lifelong lifestyle changes. Patients are encouraged to choose a low-fat diet, which aids weight reduction in the overweight patient and establishes healthy eating habits in all patients. Because alcohol can exacerbate symptoms, diabetics are encouraged to limit consumption. Diabetics must keep close tabs on the amount of carbohydrates (sugars and starches) they eat and are advised to avoid simple sugars such as those found in candy and desserts. Exercise is another must for the diabetic (as it is for everyone). Regular exercise, even of moderate intensity (such as brisk walking), improves insulin sensitivity and reduces the risk of heart disease in people with type 2 diabetes, even if they have no cardiac risk factors other than diabetes. In fact, studies of older people who engage in regular to moderate aerobic exercise show they lower their risk for diabetes, even if they don't lose weight.

Emerging studies suggest that certain alternative therapies may benefit the diabetic patient. Among the treatments that have been studied are acupuncture, biofeedback, guided imagery, and vitamin and mineral supplementation. It is important to note that many alternative treatments remain either untested or unproven through traditional scientific studies. Acupuncture is a procedure in which a practitioner sticks needles into designated pressure points on the body. Some Western scientists believe that acupuncture triggers the release of the body's natural painkillers. Diabetics suffering from neuropathy may derive some benefit from this therapy.

Biofeedback is a technique that helps patients become more aware of and learn to deal with the body's response to pain. This alternative therapy emphasizes relaxation and stress-reduction techniques. Reducing stress keeps the adrenal gland from going into overdrive and triggering a cascade of endocrine responses. Research has shown that stress can increase a diabetic's glucose level.

Guided imagery is a relaxation technique that is usually performed by a biofeedback practioner. During this procedure, a person thinks of peaceful mental images, such as ocean waves. A person may also focus on controlling or curing a chronic disease, such as diabetes.

The benefit of chromium for diabetes has been studied and debated for years. The body uses chromium to make glucose tolerance factor, which helps improve insulin action. Several studies report that chromium supplementation may improve diabetes control; however, due to insufficient information on its use in diabetes, no recommendations for supplementation yet exist.

Magnesium is another substance that has fueled debate in the endocrinology community. Although the relationship between magnesium and diabetes has been studied for decades, it is not yet fully understood. Studies suggest that magnesium deficiencies worsen the blood sugar control in type 2 diabetes. Researchers theorize that the deficiency disrupts insulin production in the pancreas and increases insulin resistance in the body's tissues. Evidence suggests that a deficiency of magnesium may contribute to certain diabetes complications.

Vanadium is a compound present in microscopic levels in plants and animals. One study has found that when people with diabetes took vanadium, they developed a modest increase in insulin sensitivity and consequently could decrease their insulin requirements. Currently researchers are researching vanadium's mechanism in the body and investigating potential side effects and dose efficacy.

Preventing Diabetes

Research studies conducted in the United States and abroad have found that lifestyle changes (such as a healthy diet and moderate exercise) can prevent or delay the onset of type 2 diabetes among high-risk adults. These studies included people with IGT and other high-risk characteristics for developing diabetes. Studies also suggest that certain medications can prevent diabetes in some population groups. The Diabetes Prevention Program (DPP) was a major clinical trial comparing diet and exercise to treatment with metformin in more than 3,000 people with impaired glucose tolerance, a condition that often precedes diabetes. Study investigators found that those who took the drug metformin reduced their risk of getting type 2 diabetes by 31 percent. Treatment with the drug was most effective among people 25–40 years of age who were 50 to 80 pounds overweight.

ISLET CELL TUMORS OF THE PANCREAS

Islet cell cancer is a rare disease in which malignant tumors are found in certain tissues of the endocrine pancreas. Symptoms include diarrhea, stomach pain, persistent fatigue, fainting, or gaining weight without eating too much. Different tumors can also cause an array of effects. There are several types of islet cell tumors.

Gastrinoma, or Zollinger-Ellison Syndrome, is a tumor that produces an excess of the hormone gastrin. Normally, gastrin stimulates the release of gastric acid into the stomach cavity to aid digestion (see Chapter 5). When too much stomach acid is present, ulcers may develop. Diagnosis is confirmed by the presence of excess gastrin in the blood and stomach acid.

An insulinoma is a pancreatic tumor that gives off too much insulin, causing the body to store sugar instead of burning it for energy. This leads to low blood sugar, or hypoglycemia. Symptoms include weakness, trembling, dizziness, and cold perspiration. The skin appears pale or ashen. Patients may also complain of anxiety, constant hunger, blurred vision, and a tingling sensation in their extremities. Symptoms develop gradually and often occur with exercise or on an empty stomach.

Glucagonoma is a rare type of islet cell tumor that secretes too much glucagon. The keynote symptom for this tumor is a distinctive skin rash called necrolytic migratory erythema. Other symptoms include a sore tongue and weight loss. Glucagonomas are usually malignant.

Other types of islet cell cancer can affect the pancreas and/or small intestine. Each type of tumor may affect different hormones in the body and cause different symptoms, such as watery diarrhea, weight loss, and a low blood potassium level. Surgical removal of the affected tissue/tumor is the key treatment for all pancreatic tumors.

Acronyms

ACTH   Adrenocorticotropic hormone (corticotropin)

ADH   Antidiuretic hormone (vasopressin)

AI   Adrenocorticol insufficiency

ANP   Atrial natriuretic peptide

ATP   Adenosine triphosphate

BMR   Basal metabolic rate

BNP   Brain natriuretic peptide

CAH   Congenital adrenal hyperplasia

cAMP   Cyclic adenosine monophosphate

CBG   Corticosteroid-binding globulin

CCK   Cholecystokinin

CEA   Carcinoembryonic antigen

CRH   Corticotropin-releasing hormone

CT   Computed tomography

DHEA   Dehydroepiandrosterone

DHEA-S   Dehydroepiandroterone sulfate

DHT   Dihydrotestosterone

DI   Diabetes insipidus

DPP   Diabetes Prevention Program

EGF   Epidermal growth factor

EPO   Erythropoietin

EPT   Estrogen-progestin therapy

ERT   Estrogen replacement therapy

FDA   Food and Drug Administration

FSH   Follicle-stimulating hormone

GDP   Guanine diphosphate

GH   Growth hormone

GHIH   Growth hormone–inhibiting hormone

GHRH   Growth hormone–releasing hormone

GIP   Gastric inhibitory peptide

GK   Gluckokinase

GnRH   Gonadotropin-releasing hormone

GTP   Guanine triphosphalt

hCG   Human chorionic gonadotropin

HDL   High-density lipoprotein

HPA   Hypothalamic-pituitary axis

hPL   Human placental lactogen

HRT   Hormone replacement therapy

IBS   Irritable bowel syndrome

IFG   Impaired fasting glucose

IGF   Insulin-like growth factor (also IGF-1)

IGT   Impaired glucose tolerance

JAMA   *Journal of the American Medical Association*

JCEM   *Journal of Clinical Endocrinology & Metabolism*

LDL   Low-density lipoprotein

LH   Luteinizing hormone

MEN 1   Multiple endocrine neoplasia syndrome 1

MIS   Müllerian-inhibiting substance

MRI   Magnetic resonance imaging

mRNA   Messenger ribonucleic acid

MTC   Medullary thyroid carcinoma

NCEP   National Cholesterol Education Program

NEJM   *New England Journal of Medicine*

NIDD   Noninsulin-dependent diabetes

NIDDK   National Institute of Diabetes and Digestive and Kidney Diseases

OGTT	Oral glucose tolerance test	**SHBG**	Sex hormone–binding globulin
PDGF	Platelet-derived growth factor	**SRY**	Sex-determining region
PP	Pancreatic polypeptide	**TB**	Tuberculosis
PRL	Prolactin	**TBG**	T_4-binding globulin
PTH	Parathyroid hormone (parathormone)	**TGF**	Transforming growth factor
PTHrP	Parathyroid hormone–related protein	**TNF**	Tumor necrosis factor
PTU	Propylthiouracil	**TRH**	Thyrotropin-releasing hormone
RCC	Rathke's cleft cysts	**TSH**	Thyroid-stimulating hormone
RDS	Respiratory Distress Syndrome	**VIP**	Vasoactive intestinal polypeptide
SERM	Selective estrogen receptor modulator	**WHI**	Women's Health Initiative

Glossary

Acetylcholine Neurotransmitter that stimulates or inhibits muscle contractions.

Acromegaly A hormonal disorder caused by excess growth hormone (GH) production by the pituitary gland.

Adipose tissue Storage site for fats.

Adrenarche The part of puberty triggered by adrenal androgens, characterized by a growth spurt and body hair growth.

Adrenergic receptors Cellular receptors for adrenal hormones.

Adrenocorticotropic hormone (ACTH) Hormone produced by the pituitary gland, which stimulates the release of hormones from the adrenal cortex. Also called corticotropin.

Agonists Hormones that bind to their receptor and elicit a specific biological response.

Aldosterone Steroid hormone produced by the adrenal cortex that helps regulate sodium and potassium balance.

Alveoli Milk-producing cells in the breasts.

Amennorhea The absence or stopping of the menstrual cycle.

Amino acids Organic compounds that form the building blocks of proteins.

Androgens Male sex hormones produced by the gonads and adrenal cortex.

Androstenedione Sex steroid produced in the adrenal cortex that can be converted into estrogen and testosterone.

Angiotensin I Hormone produced from angiotensinogen in the liver under the stimulation of the peptide, renin.

Angiotensin II Hormone produced from angiotensin I that constricts the blood vessels and stimulates the adrenal cortex to produce aldosterone.

Angiotensinogen Amino acid from the liver that is converted by the enzyme renin into angiotensin I.

Antagonists Hormones that bind to the receptor but do not trigger a biological response. By occupying the receptor, an antagonist blocks an agonist from binding and thus prevents the triggering of the desired effect within the cell.

Anterior pituitary The lobe of the pituitary that secretes hormones that stimulate the adrenal glands, thyroid gland, ovaries, and testes.

Antidiuretic hormone (ADH) Hormone produced by the pituitary gland that increases the permeability of the kidney ducts to return more fluid to the bloodstream. Also called vasopressin.

Atrial natriuretic peptide (ANP) Hormone produced by muscle cells in the heart that regulates salt and water balance and blood pressure.

Atrophy A decrease in size or wasting away of a body part or tissue.

Autocrine The action of a hormone on the cells that produced it.

Basal metabolic rate (BMR) The rate at which heat is given off by the body at complete rest.

Brain natriuretic peptide (BNP) Hormone produced by muscle cells in the heart that regulates salt and water balance and blood pressure. This hormone is related to atrial natriuretic peptide (ANP).

Calciferol Vitamin D_2, a fat-soluble vitamin that regulates blood levels of calcium and phosphorous.

Calcitonin Hormone produced by the thyroid gland that influences calcium and phosphorous levels in the blood.

Carcinoembryonic antigen (CEA) A protein present in fetal stomach tissues during the first two trimesters of pregnancy. It is also found in the blood of patients with some forms of cancer.

Catecholamines A class of hormone (including epinephrine and norepinephrine) synthesized in the adrenal medulla that is involved in the body's stress response.

Cholecalciferol Vitamin D_3, which is generated in the skin through exposure to sunlight.

Cholecystokinin Hormone produced by the gastrointestinal system that triggers the release of digestive enzymes in the small intestine and stimulates gallbladder contraction.

Chondrocytes Cartilage cells.

Conglobate glands Glands that form into a round compact mass.

Conglomerate glands Glands that collect together into a ball or a mass.

Corpus luteum Progesterone-secreting tissue that forms from a ruptured Graafian follicle in the mammalian ovary after the egg has been released.

Corticotroph Cell in the anterior pituitary gland that secretes corticotropin (ACTH).

Corticotropin Also called adrenocorticotropic hormone (ACTH). Hormone produced by the pituitary gland that stimulates the release of hormones from the adrenal cortex.

Corticotropin-releasing hormone (CRH) Peptide hormone released by the hypothalamus that stimulates the synthesis of adrenocorticotropic hormone (ACTH) from the pituitary gland.

Cortisol A steroid hormone produced by the adrenal cortex that influences glucose metabolism. Also called hydrocortisone.

Cretinism A rare congenital disorder characterized by physical stunting and mental retardation; caused by severe hypothyroidism.

Cytokines Signaling peptides secreted by immune cells and other types of cells in response to infection or other stimuli.

Dehydroepiandrosterone (DHEA) Weak sex steroid produced by the adrenal cortex that can be converted into testosterone and estrogen.

Dihydrotestosterone (DHT) A potent male hormone converted from testosterone.

Dopamine Neurotransmitter that influences the brain processes that control pain, pleasure, emotion, and movement.

Dwarfism An exaggerated slow growth pattern in childhood resulting from the underproduction of growth hormone.

Effector A molecule that regulates a series of chemical reactions.

Eicosanoids Compounds derived from fatty acids that act like hormones to influence physiologic functions.

Electrolytes Electrically charged chemical ions such as sodium, potassium, chloride, calcium, magnesium, and phosphate.

Epinephrine A catecholamine produced by the adrenal medulla that is involved in the body's stress response.

Epiphyses Plates in the long bones. When they close following puberty, bone growth ceases.

Erythropoietin Protein hormone produced by the kidneys that stimulates red blood cell production.

Estradiol The most potent of the three forms of estrogen, estradiol influences maturation of the female sexual organs and is involved in body fat placement.

Estriol A weak form of estrogen.

Estrogen Any of a family of hormones produced by the female ovaries that determine female sexual characteristics and influence reproductive development.

Estrone The weakest of the three forms of estrogen.

Exocrine Producing, being, or relating to a secretion that is released outside its source, for example exocrine pancreatic cells.

Exophthalmos A protrusion of the eyeballs in their sockets. This symptom is commonly found in patients with Graves' disease.

Extracellular fluid Fluid located in the spaces between cells.

Extranodal lymphomas A tumor of lymphoid tissue that arises outside the lymph nodes.

Fibroblasts Cells that make up connective tissue.

Follicle-stimulating hormone (FSH) A hormone produced by the anterior pituitary gland that triggers sperm production in the testes and stimulates the development of follicles in the ovaries.

Ganglioneuromatosis The condition of having many widespread, slow-growing tumors in the brain or spinal cord.

Gastric inhibitory peptide (GIP) Gastrointestinal hormone whose main action is to block the secretion of gastric acid.

Gastrin Hormone produced by the gastrointestinal system that regulates stomach acid secretion.

Gene transcription The process by which a strand of DNA is copied to form a complementary RNA strand.

Ghrelin A peptide hormone secreted by the stomach that stimulates growth hormone secretion.

Gigantism An exaggerated growth period most commonly caused by over excretion of growth hormone during childhood.

Glucagon Hormone produced by the endocrine pancreas that increases blood glucose levels.

Glucocorticoids A class of hormones synthesized by the adrenal cortex that regulate glucose metabolism.

Gluconeogenesis The production of glucose from noncarbohydrate sources in the liver between meals.

Glycerol An alcohol used in metabolism.

Glycogen The stored form of glucose.

Glycogenolysis The breakdown of glycogen in the liver and in muscle tissue.

Glycoprotein An organic compound composed of a joined protein and carbohydrate.

Goiter A benign enlargement of the thyroid gland that is visible as a swelling in the front of the neck.

Gonadocorticoids Sex hormones (androgen and estrogen) produced by the adrenal cortex that influence the development of male and female characteristics.

Gonadotroph Cell in the anterior pituitary gland that secretes luteinizing hormone and follicle-stimulating hormone.

Gonadotropins Hormones (luteinizing hormone and follicle-stimulating hormone) released by the anterior pituitary gland that stimulate the ovaries and testes.

Gonadotropin-releasing hormone (GnRH) Hypothalamic neurohormone that stimulates the release of luteinizing hormone (LH) and follicle-stimulating hormone (FSH) from the pituitary gland.

Graafian follicle A liquid-filled cavity in the mammalian ovary that releases a mature egg before ovulation.

Growth factors Proteins that act on cells to stimulate differentiation and proliferation.

Growth hormone Hormone secreted by the anterior pituitary gland that promotes bone and muscle growth and metabolism.

Growth hormone-releasing hormone (GHRH) Peptide hormone released by the hypothalamus that stimulates the secretion of growth hormone from the pituitary gland.

Hashimoto's thyroiditis An autoimmune condition in which the body's immune system attacks the thyroid gland.

Hemochromatosis A metabolic disorder that occurs from the deposition of iron-containing pigments in the tissues and is marked by bronzing of the skin, diabetes, and weakness.

High-density lipoprotein (HDL) Otherwise known as "good cholesterol," HDL transports cholesterol from peripheral tissues to the liver to prevent a build-up in blood vessels.

Histamine Neurotransmitter involved in the immune response and allergic reactions.

Histiocytosis X A generic name for a group of syndromes characterized by an abnormal increase in the number of certain immune cells including monocytes, macrophages, and dendritic cells.

Homeostasis The regulation of the body's internal environment to maintain balance.

Hormone replacement therapy (HRT) Replacing estrogen and progestrone to relieve the symptoms of menopause.

Hormones Chemical substances produced by tissues in one part of the body that travel through the bloodstream or act locally on other cells and tissues.

Human chorionic gonadotropin (hCG) Protein produced by the fetal placenta that regulates steroid hormone synthesis and helps maintain the pregnancy.

Human placental lactogen (hPL) Protein hormone, similar to prolactin and growth hormone, that stimulates breast growth and development and that alters maternal glucose metabolism to free up additional glucose for the fetus.

Hypercalcemia Too much calcium in the blood.

Hyperglycemia High blood glucose.

Hyperthyroidism A condition that occurs when levels of thyroid hormone in the blood are excessively high.

Hypocalcemia A deficiency of calcium in the blood.

Hypoglycemia Low blood glucose.

Hypophysis The pituitary gland.

Hypothalamic-hypophyseal portal system Circulation system through which neurohormones from the hypothalamus travel directly to the anterior pituitary gland without ever entering the general circulation.

Hypothalamic-pituitary-target organ axis Multiloop feedback system that coordinates the efforts of the hypothalamus, the pituitary gland, and the target gland.

Hypothalamus Region of the brain that coordinates the neuroendocrine system and releases neurohormones that signal the pituitary gland to release its hormones.

Hypothyroidism A condition caused by a deficiency of thyroid hormones in the blood.

Hysterectomy Surgical removal of the uterus.

Idiocy Extreme mental retardation, commonly due to incomplete or abnormal brain development.

Inhibin Hormone secreted by the ovaries and testes that inhibits the release of follicle-stimulating hormone (FSH) by the pituitary.

Insulin A hormone produced by the endocrine pancreas that controls glucose levels in the blood by influencing carbohydrate, fat, and protein metabolism.

Insulin-like growth factors Substances produced in the liver and other tissues that act much like growth hormone, stimulating bone, cartilage, and muscle cell growth and differentiation.

Intermediate pituitary A lobe of the pituitary of which only vestiges remain in humans.

Intracellular fluid Fluid located within the cells.

Islets of Langerhans Endocrine cells located in the pancreas in which the hormones insulin and glucagon are produced.

Ketone bodies Substances produced from fats when not enough glucose is present, which provide an alternate energy source for the brain and other tissues.

Lactotroph Cell in the anterior pituitary gland that secretes prolactin.

Leptin A protein hormone that influences metabolism and regulates body fat.

Leydig (interstitial) cells Testosterone-producing cells of the testes.

Lipodystrophy A clinical condition characterized by a poor or uneven distribution of fat cells that can lead to large amounts of fat storage in inappropriate places. The syndrome can cause lower belly obesity and a buffalo-like hump on the upper back.

Lipolysis The breakdown and release of fatty acids from fat cells into the blood.

Lipoprotein Protein made up of lipids and fats that transports cholesterol and fatty acids through the bloodstream.

Low-density lipoproteins (LDLs) Lipid-protein complexes that transport cholesterol to tissues; otherwise known as "bad cholesterol."

Luteinizing hormone (LH) A hormone produced and secreted by the anterior pituitary gland that stimulates ovulation and menstruation in women and androgen synthesis by the testes in men.

Luteolysis The process by which the corpus luteum in the ovary degenerates when an egg is not fertilized.

Megacolon (toxic) A life-threatening complication of certain intestinal conditions, marked by a very dilated colon, a swollen abdomen, and sometimes fever, stomach pain, or shock.

Melatonin Hormone secreted by the pineal gland that responds to light and darkness and that influences the body's biological clock.

Menarche The first menstruation.

Mineralocorticoids A class of hormones produced by the adrenal cortex that regulate mineral metabolism.

Motilin Gastrointestinal hormone that stimulates intestinal muscle contractions to clean undigested materials from the small intestine.

Myxedema A severe form of hypothyroidism marked by firm, inelastic edema; dry skin and hair; and loss of mental and physical vigor. Myxedema coma is life-threatening.

Neuroendocrine system Collective term for the endocrine glands and the parts of the nervous system that regulate endocrine function.

Neuroendocrinology The study of the neuroendocrine system.

Neurohormone A chemical messenger released by the hypothalamus that signals the pituitary gland to release or inhibit release of its hormones.

Neuron A cell of the nervous system.

Neurosecretory cells Specialized nerve cells that transmit chemical impulses, release hormones, and serve as a link between the endocrine and nervous systems.

Neurotransmitters Chemical messengers that are produced and released by the nervous system and are used by nerve cells to communicate.

Norepinephrine A catecholamine produced by the adrenal medulla that is involved in the body's stress response.

Oogenesis The formation and development of an egg in the ovary.

Osmoreceptors Neurons that sense fluid concentrations and send a message to the hypothalamus.

Osteoblasts Bone-building cells.

Osteoclasts Bone-removing cells.

Osteoporosis A condition characterized by a decrease in bone mass.

Oxytocin Hormone synthesized by the pituitary gland that stimulates uterine contractions during labor and triggers milk let-down during lactation.

Pancreatic polypeptide Hormone secreted by the F cells of the endocrine pancreas that inhibits gallbladder contraction and halts enzyme secretion by exocrine cells in the pancreas.

Panhypopituitarism Syndrome characterized by the complete failure of the pituitary gland to produce its hormones.

Paracrine The action of a hormone on neighboring cells.

Parathormone See parathyroid hormone.

Parathyroid hormone (PTH) A hormone secreted by the parathyroid gland that helps maintain calcium and phosphorous levels in the body. PTH controls the release of calcium from bone, the absorption of calcium in the intestine, and the excretion of calcium in the urine. Also called parathormone.

Phenochromocytoma An uncommon adrenaline-secreting tumor.

Pinealocytes Cells in the pineal gland that produce the hormone melatonin.

Pitocin A synthetic form of the hormone oxytocin that is used to induce labor.

Placental lactogen Hormone produced by the fetal placenta that influences metabolism and growth.

Polymorphism A condition in which a chromosome or other genetic material occurs in more than one form.

Polyunsaturated fatty acids Components of dietary fats that contain at least two double bonds.

Posterior pituitary Lobe of the pituitary gland that is an extension of the nervous system.

Pregnenolone A steroid hormone precursor produced from cholesterol.

Preprohormone/prohormone An inactive sequence of amino acids from which an active hormone is released.

Pretibial myxedema A rare complication of Graves' disease marked by diffuse, nonpitting edema and thickening of the skin, usually on the anterior aspect of the lower legs and spreading to the upper surface of the feet.

Progesterone Steroid hormone produced in the adrenal gland, placenta, and corpus luteum that influences sexual development and reproduction.

Progestin Female hormone produced by the ovaries that influences sexual development and pregnancy.

Proglucagon Precursor molecule from which the hormone glucagon is produced.

Proinsulin The inactive precursor molecule from which insulin is formed.

Prolactin A protein hormone secreted by the anterior pituitary that stimulates mammary gland development and milk production.

Prostaglandin Fatty acid derivatives that act much like hormones to influence a number of physiological processes throughout the body.

Receptors Proteins on the surface of cells or within cells that bind to particular hormones.

Relaxin Ovarian peptide hormone that relaxes the pelvic ligaments and softens the cervix to ease childbirth.

Renin An enzyme released by the kidneys that converts the amino acid angiotensinogen to angiotensin I.

Sarcoidosis A chronic disease of unknown cause marked by the formation of nodules, especially in the lymph nodes, lungs, bones, and skin.

Secretin Hormone produced by the gastrointestinal system that stimulates pancreatic secretion during digestion.

Seminiferous tubules Tubes in the testes in which sperm are produced.

Serotonin Neurotransmitter that influences sleep, mood, appetite, and muscle contractions.

Sertoli cells Cells in the testes in which sperm is produced.

Sex hormone–binding globulin (SHBG) Protein that carries the sex hormones through the bloodstream to their target cells.

Somatomedin (Insulin-like growth factor 1) Growth-hormone-like substance produced in the liver and in many other tissues that promotes bone, muscle, and cartilage growth.

Somatostatin Hormone produced by the endocrine pancreas and hypothalamus that regulates insulin and glucagon release, and inhibits growth hormone release from the pituitary gland.

Somatotroph Cell in the anterior pituitary gland that secretes growth hormone.

Spermatogenesis Sperm production.

Substance P Neuropeptide found in the gut and brain that stimulates smooth muscle contraction and epithelial cell growth and that plays a role in both the pain and pleasure responses.

Synapse The junction between two nerve cells across which nerve impulses pass.

Tachycardia A rapid heart rate.

Target cells Cells that are responsive to a particular hormone.

Testosterone A steroid hormone produced mainly in the testes that influences male sexual development.

Thelarche Breast growth during puberty.

Thyroglobulin Glycoprotein from which the thyroid hormones, thyroxine (T_4) and triiodothyronine (T_3), are produced.

Thyroid-stimulating hormone (TSH) Hormone produced by the pituitary gland that stimulates the thyroid gland to secrete its hormones, thyroxine (T_4) and triiodothyronine (T_3). Also called thyrotropin.

Thyrotroph Cell in the anterior pituitary gland that secretes thyroid-stimulating hormone.

Thyrotropin See thyroid-stimulating hormone.

Thyrotropin-releasing hormone (TRH) Hypothalamic neurohormone that triggers the release of thyroid-stimulating hormone (TSH) and prolactin (PRL) from the pituitary gland.

Thyroxine (T_4) Thyroid hormone that influences metabolism and growth.

Triglycerides Fatty compounds that circulate in the blood and are stored in tissues.

Triiodothyronine (T_3) The more potent of the two thyroid hormones.

Tropic hormones Hormones that control the release of other hormones.

Tyrosine An amino acid component of protein.

Umbilical hernia A small protrusion in the infant abdominal wall near the navel.

Vasoactive intestinal polypeptide (VIP) A polypeptide secreted by cells in the intestinal tract that influences intestinal electrolyte and water secretion.

Vasopressin Hormone produced by the pituitary gland that increases the permeability of the kidney ducts to return more fluid to the bloodstream. Also called antidiuretic hormone (ADH).

Zona fasciculata The middle layer of the adrenal cortex, in which the glucocorticoids (cortisol) are produced.

Zona glomerulosa The outermost layer of the adrenal cortex, in which the mineralocorticoids (aldosterone) are produced.

Zona reticularis The innermost layer of the adrenal cortex, in which the gonadocorticoids (sex hormones) are produced.

Organizations and Web Sites

American Association of Clinical Endocrinologists
1000 Riverside Avenue, Suite 205
Jacksonville, FL 32204
Phone: (904) 353-7878
www.aace.com

AACE provides and promotes education, research, and communication in the art and science of clinical endocrinology.

American Association of Endocrine Surgeons
MetroHealth Medical Center, H920
2500 MetroHealth Drive
Cleveland, OH 44109-1908
Phone: (216) 778-4753
http://endocrinesurgeons.org

This membership society is dedicated to the science and art of endocrine surgery.

American Cancer Society
2200 Century Parkway, Suite 950
Atlanta, GA 30345
Phone: (800) 282-4914
www.cancer.org

The American Cancer Society Web site provides information regarding cancer news, breakthroughs, and education.

American Diabetes Association
1701 North Beauregard Street
Alexandria, VA 22311
Phone: (800) 342-2383
www.diabetes.org

The American Diabetes Association publishes scientific findings and provides information and other services to those interested in learning about diabetes.

American Neuroendocrine Society
Dr. James Koenig, Secretary
Maryland Psychiatric Research Center
P.O. Box 21247
Baltimore, MD 21228
Phone: (410) 402-7319
www.neuroendocrine.org

The American Neuroendocrine Society functions to promote and encourage basic and applied research in all aspects of neuroendocrinology.

American Thyroid Association
6066 Leesburg Pike, Suite 650
Falls Church, VA 22041
Phone: (703) 998-8890
www.thyroid.org

This organization promotes research and education on thyroid physiology and diseases.

Endocrine Nurses Society
8401 Connecticut Avenue, Suite 900
Chevy Chase, MD 20815
Phone: (301) 941-0249
www.endo-nurses.org

The Endocrine Nurses Society is a professional organization that promotes excellence in the clinical care of patients through the advancement of the science and art of endocrine nursing.

Endocrine Society
8401 Connecticut Avenue, Suite 900
Chevy Chase, MD 20815
Phone: (301) 941-0200
www.endo-society.org

The Endocrine Society is the largest professional organization of endocrinologists in the world. Its goal is to promote research and education in the field of endocrinology.

Endocrine Web
www.endocrineweb.com

Endocrine Web provides more than 120 very detailed but easy-to-understand pages on endocrine diseases, conditions, hormone problems, and treatment options including all types of thyroid, parathyroid, and adrenal surgery.

Hormone Foundation
8401 Connecticut Avenue, Suite 900
Chevy Chase, MD 20815
Phone: (301) 951-2602
www.hormone.org

The Hormone Foundation is the public education affiliate of the Endocrine Society. It serves as a resource for the public by promoting the prevention, treatment, and cure of hormone-related conditions.

Human Growth Foundation
997 Glen Cove Avenue
Glen Head, NY 11545
Phone: (800) 451-6434
www.hgfound.org

The Human Growth Foundation provides information on growth hormone disorders.

Lawson Wilkins Pediatric Endocrine Society
http://lwpes.org

The Lawson Wilkins Pediatric Endocrine Society offers information on endocrine disorders and diseases relating to children.

National Diabetes Information Clearinghouse
1 Information Way
Bethesda, MD 20892-3560
Phone: (800) 860-8747
http://diabetes.niddk.nih.gov

The National Diabetes Information Clearinghouse (NDIC) is an information dissemination service of the National Institute of Diabetes and Digestive and Kidney Diseases (NIDDK). The NIDDK is part of the National Institutes of Health (NIH). It provides information about the latest diabetes research for patients, health care professionals, and the general public.

Nephrogenic Diabetes Insipidus Foundation
Main Street
P.O. Box 1390
Eastsound, WA 98245
Phone: (888) 376-6343
www.ndif.org

Visitors to this Web site will learn about the causes, diagnosis, and treatment of nephrogenic diabetes insipidus.

Paget Foundation for Paget's Disease of Bone and Related Disorders
120 Wall Street, Suite 1602
New York, NY 10005-4001
Phone: (800) 23-PAGET
www.paget.org

This Web site includes information for patients and health professionals on Paget's disease of bone, primary hyperparathyroidism, fibrous dysplasia, osteopetrosis (not osteoporosis), and the complications of some cancers to the bones.

Pituitary Society
New York University Medical Center
550 First Avenue
New York, NY 10016

Phone: (212) 263-6772

http://pituitarysociety.med.nyu.edu/

This organization provides information on diseases of the pituitary gland.

Society for Endocrinology
22 Apex Court
Woodlands, Bradley Stoke
Bristol BS32 4JT
United Kingdom
Phone: +44 (0) 1454 642200
www.endocrinology.org

This membership society was set up in 1946 to promote the advancement of endocrinology.

Thyroid Foundation of America
410 Stuart Street
Boston, MA 02116
Phone: (800) 832-8321
www.tsh.org

This nonprofit organization helps patients understand and deal with their thyroid disease.

Women in Endocrinology
www.women-in-endo.org

This organization is devoted to promoting and facilitating the professional development and advancement of women in the field of endocrinology.

Bibliography

Adams, F., ed. and trans. "Aretaeos the Kappadokian: The Extant Works." In Medvei, *The History of Clinical Endocrinology*, 34.

"Addison, Thomas." *Encyclopedia Britannica* from Encyclopedia Britannica Premium Service. http://www.britannica.com/eb/article?eu=3742 (accessed July 21, 2003).

"Addison's disease." NIH Publication No. 90-3054, November 1989. Also available at: United States National Institute of Diabetes and Digestive and Kidney Diseases of the National Institute of Health. http://www.niddk.nih.gov/health/endo/pubs/addison/addison.htm (accessed June 2003).

Alton Museum of History and Art. "Robert Pershing Wadlow." http://www.altonweb.com/history/wadlow/ (accessed September 2003).

"Are You at Risk for Gestational Diabetes?" NIH Publication No. 00-4818, December 2000. Also available at: National Institute of Child Health and Human Development. http://www.nichd.nih.gov/publications/pubs/gest_diabetes.htm (accessed June 2003).

"As Diabetes Epidemic Surges, HHS and ADA Join Forces to Fight Heart Disease, the Leading Cause of Death for People with Diabetes." National Diabetes Information Clearinghouse press release, November 1, 2001. http://www.niddk.nih.gov/welcome/releases/11–01–01.htm (accessed June 2003).

Austgen, Laura, B. A. Bowen, and Melissa Rouge. "Pathophysiology of the Endocrine System." Colorado State University. http://www.vivo.colostate.edu/hbooks/pathphys/endocrine/ (accessed March 22, 2003).

"Banting, Sir Frederick Grant." *Encyclopedia Britannica* from Encyclopedia Britannica Premium Service. http://www.britannica.com/eb/article?eu=13372 (accessed July 23, 2003).

"Basic Diabetes Information." American Diabetes Association online. http://www.diabetes.org (accessed June 2003).

"Bayliss, Sir William Maddock." *Encyclopedia Britannica* from Encyclopedia Britannica Premium Service. http://www.britannica.com/eb/article?eu=14059 (accessed July 23, 2003).

Beers, Mark, ed. *The Merck Manual of Medical Information, Second Edition*. White-house Station, NJ: Merck & Co., 2003.

"Bernard, Claude."*Encyclopedia Britannica* from Encyclopedia Britannica Premium Service. http://www.britannica.com/eb/article?eu=80966 (accessed July 21, 2003).

"Birth Control."*Encyclopedia Britannica* from Encyclopedia Britannica Premium Service. http://www.britannica.com/eb/article?eu=108650 (accessed July 26, 2003).

Blackburn, George L., and Laura C. Bevis. "The Obesity Epidemic: Prevention and Treatment of the Metabolic Syndrome." Medscape. http://www.medscape.com (accessed June 2003).

"Brown-Séquard, Charles-Édouard."*Encyclopedia Britannica* from Encyclopedia Britannica Premium Service. http://www.britannica.com/eb/article?eu=1844 (accessed July 21, 2003).

Bubnow, N. A. Article from *Z. Physiol. Chemie*. In Medvei, *The History of Clinical Endocrinology*, 113.

Cawley, Thomas. "A Singular Case of Diabetes, Consisting Entirely in the Quality of the Urine; with an Inquiry into the Different Theories of That Disease." In Medvei, *The History of Clinical Endocrinology*, 975.

Chatin, G. A. "Recherche comparative de l'iode ed de quelques autres principes dans les eaux (et les egouts) qui alimentent Paris, Londres et Turin." In Medvei, *The History of Clinical Endocrinology*, 138–139.

Collins, Angela Smith. "More than a Pump: The Endocrine Functions of the Heart." *Journal of Critical Care* 10 (2001). http://www.aacn.org/AACN/jrnlajcc.nsf/GetArticle/ArticleFour102?OpenDocument (accessed June 2003).

"Congenital Adrenal Hyperplasia." MEDLINEplus Medical Encyclopedia. http://www.nlm.nih.gov/medlineplus/ency/article/000411.htm (accessed June 2003).

"Courtois, Bernard." *Encyclopedia Britannica* from Encyclopedia Britannica Premium Service. http://www.britannica.com/eb/article?eu=27056 (accessed July 21, 2003).

"Diabetes." National Diabetes Information Clearinghouse, NIH Publication No. 03-3873, May 2003. Also available at: United States National Institute of Diabetes and Digestive and Kidney Diseases of the National Institute of Health. http://diabetes.niddk.nih.gov/dm/pubs/overview/index.htm (accessed June 2003).

"Diet and Exercise Dramatically Delay Type 2 Diabetes: Diabetes Medication Metformin Also Effective." National Institute of Diabetes and Digestive and Kidney Diseases. http://www.niddk.nih.gov/welcome/releases/8_8_01.htm (accessed June 2003).

"Doisy, Edward Adelbert."*Encyclopedia Britannica* from Encyclopedia Britannica Premium Service. http://www.britannica.com/eb/article?eu=31312 (accessed July 23, 2003).

"Endocrine Disrupters." Environmental Protection Agency. http://www.epa.gov/scipoly/oscpendo/ (accessed June 2003).

"Endocrine Disrupters." The Why Files. http://whyfiles.org/045env_hormone/main2.html (accessed March 2004).

"Endocrine System, Human." *Encyclopedia Britannica* from Encyclopedia Britannica Premium Service. http://www.britannica.com/eb/article?eu=108523 (accessed April 10, 2003).

"Endocrinology." *Grolier Encyclopedia*. New York: Grolier Incorporated, 2001: 339.

"Endocrinology Health Guide." University of Maryland. http://www.umm.edu/endocrin/. (accessed March 29, 2003).

"Facts about Postmenopausal Hormone Therapy." NIH Publication No. 02-5200, October 2002. Also available at: National Heart, Lung, and Blood Institute. http://www.nhlbi.nih.gov/health/women/pht_facts.htm (accessed August 2003).

"FDA Approves New Labeling and Provides New Advice to Postmenopausal Women Who Use or Who Are Considering Using Estrogen and Estrogen with Progestin." U.S. Food and Drug Administration Fact Sheet, January 8, 2003. http://www.fda.gov/oc/factsheets/WHI.html (accessed August 2003).

"Former President Bush 'Fine,' Leaves Hospital." CNN.com, February 25, 2000. http://www.cnn.com/2000/US/02/25/bush.hospitalized.03/ (accessed September 2003).

Frank, A.E. "Ueber Beziehungen der Hypophyse zum Diabetes insipidus." In Medvei, *The History of Clinical Endocrinology,* 232.

Frank, J.P. "De curandis hominum morbis epitome." In Medvei, *The History of Clinical Endocrinology,* 97.

"From Elixirs to Sitcoms." *U.S. News & World Report* (November 18, 2002), 44.

Garrison, F.H. "An Introduction to the History of Medicine." In Medvei, *The History of Clinical Endocrinology,* 65.

Graves, Robert J. Article from *London Medical and Surgical Journal. (Renshaw's)* 7. In Medvei, *The History of Clinical Endocrinology,* 142.

"Graves, Robert James."*Encyclopedia Britannica* from Encyclopedia Britannica Premium Service. http://www.britannica.com/eb/article?eu=38546 (accessed July 21, 2003).

Greenspan, Francis S., and David G. Gardner. *Basic and Clinical Endocrinology,* 6th ed. New York: Lange Medical Books/McGraw-Hill, 2001.

Griffen, James E., and Sergio R. Ojeda, eds. *Textbook of Endocrine Physiology,* 4th ed. New York: Oxford University Press, 2000.

"Haller, Albrecht von."*Encyclopedia Britannica* from Encyclopedia Britannica Premium Service. http://www.britannica.com/eb/article?eu=39741 (accessed July 17, 2003).

Hench, P.S., E.C. Kendall, et al. Article from *Mayo Clinic Proceedings* 24. In Medvei, *The History of Clinical Endocrinology,* 289.

Hertz, S., A. Roberts, and R.D. Evans. "Radioactive Iodine as an Indicator in the Study of Thyroid Physiology." In Medvei, *The History of Clinical Endocrinology,* 313.

"Hormone Replacement Therapy: Weighing the Pros and Cons." *PDR Family Guide to Women's Health & Prescription Drugs,* Annual (2001). Also available at: http://www.pdrhealth.com/content/women_health/chapters/fgwh31.shtml.

"Hormone Replacement Therapy: What to Do." MayoClinic.com. http://www.mayoclinic.com/invoke.cfm?id=WO00037 (accessed August 2003).

Hulley, S., C. Furberg, E. Barrett-Connor, J. Cauley, D. Grady, W. Haskell, et al. "Noncardiovascular Disease Outcomes during 6.8 Years of Hormone Therapy." *JAMA* 288 (2002): 58–66.

Iason, A.H. "The Thyroid Gland in Medical History." In Medvei, *The History of Clinical Endocrinology,* 88.

"Introduction to Diabetes." National Diabetes Information Clearinghouse (NDIC): A Service of the National Institute of Diabetes and Digestive and Kidney Diseases (NIDDK), NIH. http://diabetes.niddk.nih.gov/intro/index.htm (accessed February 29, 2004).

"Introduction to the Endocrine System." National Cancer Institute. http://training.seer .cancer.gov/module_anatomy/unit6_1_endo_intro.html (accessed April 21, 2003).

Jedrassik, E. Article from *Arch. Psychiatrie* 17. In Medvei, *The History of Clinical Endocrinology*, 141.

"Kocher, Emil Theodor."*Encyclopedia Britannica* from Encyclopedia Britannica Premium Service. http://www.britannica.com/eb/article?eu=46923 (accessed July 21, 2003).

Kopin, Alan S., et al. "The Cholecystokinin-A Receptor Mediates Inhibition of Food Intake Yet Is Not Essential for the Maintenance of Body Weight." *Journal of Clinical Investigation* 103 (1999): 383–391.

Larson, David E. *Mayo Clinic Family Healthbook*. New York: William Morrow and Company, 1990, 709.

Manson, A. "Medical Researches on the Effects of Iodine in Bronchocele, Paralysis, Chorea, Scrofula etc." In Medvei, *The History of Clinical Endocrinology*, 106.

"Marie, Pierre."*Encyclopedia Britannica* from Encyclopedia Britannica Premium Service. http://www.britannica.com/eb/article?eu=52162 (accessed July 21, 2003).

Matthiesen, A. F., et al. "Postpartum Maternal Oxytocin Release by Newborn: Effects of Infant Hand Massage and Suckling." *Birth* 29 (2001): 13–19.

Medvei, Victor Cornelius. *The History of Clinical Endocrinology: A Comprehensive Account of Endocrinology from Earliest Times to the Present Day*. Pearl River, NY: Parthenon, 1993.

"Menopausal Hormone Use: Questions and Answers." National Cancer Institute, July 16, 2002. Updated June 24, 2003. http://www.cancer.gov/newscenter/estro genplus (accessed August 2003).

"Menopause: An Update, 2003." National Institute on Aging. http://www.nia.nih .gov/menopause2003/index.htm (accessed August 2003).

"Menopause Core Curriculum Study Guide." North American Menopause Society. http://www.menopause.org/edumaterials/studyguide/sgtoc.html   (accessed June 2003).

"Minkowski, Oskar." *Encyclopedia Britannica* from Encyclopedia Britannica Premium Service. http://www.britannica.com/eb/article?eu=54182 (accessed July 21, 2003).

Müller, Johannes. "Handbuch der Physiologie der Menschen." In Medvei, *The History of Clinical Endocrinology*, 123.

Nachum, Zohor, et al. "Graves Disease in Pregnancy: Prospective Evaluation of a Selective Invasive Treatment Protocol." *American Journal of Obstetrics and Gynocology* 189 (July 2003): 159–165.

Neal, J. Matthew. *How the Endocrine System Works*. Malden, MA: Blackwell Science, 2002.

Oliver, G., and E. A. Schaefer. "On the physiological action of extracts of pituitary body and certain other glandular organs." In Medvei, *The History of Clinical Endocrinology*, 232.

Opie, E. L. Article from *The Journal of Experimental Medicine* 5. In Medvei, *The History of Clinical Endocrinology*, 163.

Patti, M. E., A. J. Butte, et al. "Coordinated Reduction of Genes of Oxidative Metabolism in Humans with Insulin Resistance and Diabetes: Potential Role of *PGC1* and *NRF1*." Proceedings of the National Academy of Sciences 2003, 100 (July 8, 2003): 8466–8471.

Pizzi, Richard A. "Rosalyn Yalow: Assaying the Unknown." *Modern Drug Discovery* 4, no. 9 (September 2001): 63–64. http://pubs.acs.org/subscribe/journals/mdd/v04/i09/html/09timeline.html (accessed June 2003).

Porterfield, Susan P. *Endocrine Physiology.* St. Louis, MO: Mosby, 2001.

"Postmenopausal Hormone Therapy." National Heart, Lung, and Blood Institute. http://www.nhlbi.nih.gov/health/women/q_a.htm (accessed August 2003).

"Questions and Answers on Hormone Replacement Therapy." The American College of Obstetricians and Gynecologists, August 2002. http://www.acog.org/from_home/publications/press_releases/nr08-30-02.cfm (accessed August 2003).

Rolleston, Sir H. D. "The Endocrine Organs in Health and Disease." In Medvei, *The History of Clinical Endocrinology,* 69.

Rotenstein, Deborah, MD. "Thyroid Division: Congenital Hypothyroidism." The MAGIC Foundation. http://www.magicfoundation.org/congthyr.html.

"Sanger, Frederick." *Encyclopedia Britannica* from Encyclopedia Britannica Premium Service. http://www.britannica.com/eb/article?eu=67195 (accessed July 26, 2003).

Selye, H. "The Physiology and Pathology of Exposure to Stress." In Medvei, *The History of Clinical Endocrinology,* 341.

"Selye, Hans."*Encyclopedia Britannica* from Encyclopedia Britannica Premium Service. http://www.britannica.com/eb/article?eu=1508 (accessed July 26, 2003).

Setchell, B. P. "The Contributions of Regnier de Graaf to Reproductive Biology." In Medvei, *The History of Clinical Endocrinology,* 70.

Shumaker, Sally A., Claudine Legault, Stephen R. Rapp, Leon Thal, Robert B. Wallace, Judith K. Ockene, Susan L. Hendrix, Beverly N. Jones, III, Annlouise R. Assaf, Rebecca D. Jackson, Jane Morley Kotchen, Sylvia Wassertheil-Smoller, and Jean Wactawski-Wende. "Estrogen plus Progestin and the Incidence of Dementia and Mild Cognitive Impairment in Postmenopausal Women: The Women's Health Initiative Memory Study: A Randomized Controlled Trial." *JAMA* 289 (2003): 2651–2662.

Smith, Stephen. "Hormone Therapy's Rise and Fall," *The Boston Globe,* July 20, 2003.

"Spallanzani, Lazzaro." *Encyclopedia Britannica* from Encyclopedia Britannica Premium Service. http://www.britannica.com/eb/article?eu=70767 (accessed July 18, 2003).

Spillane, J. D. "Medical Travellers." In Medvei, *The History of Clinical Endocrinology,* 95.

"Study Shows Sharp Rise in the Cost of Diabetes Nationwide." U.S. Department of Health and Human Services. http://www.hhs.gov/news/press/2003pres/20030227a.html (accessed June 2003).

"Sylvius, Franciscus." *Encyclopedia Britannica* from Encyclopedia Britannica Premium Service. http://www.britannica.com/eb/article?eu=72551 (accessed July 17, 2003).

Talmage, R. V., and J. R. Elliott. "Removal of Calcium from Bone as Influenced by the Parathyroids." In Medvei, *The History of Clinical Endocrinology,* 266.

"Thyroid Cancer–Papillary Carcinoma." U.S. National Library of Medicine .http://www.nim.nih.gov/medlineplus/ency/article/000331.htm (accessed February 28, 2004).

"To Be or Not to Be—on Hormone Replacement Therapy: A Workbook to Help You Explore Your Options." The Centers for Disease Control and Prevention, August 30, 2002.

U.S. Food and Drug Administration. "Birth Control Guide." *FDA Consumer Maga-*
zine (April 1997). http://www.fda.gov/fdac/features/1997/babytabl.html (accessed September 2003).

"What You Need to Know About Thyroid Cancer." NIH Publication No. 01-4994,
February 8, 2002, updated September 16, 2002. Also available at: National Cancer Institute. http://www.nci.nih.gov/cancerinfo/wyntk/thyroid (accessed February 28, 2004).

Index

About the Authors

STEPHANIE WATSON is an independent scholar who has written and contributed to numerous works, including *World of Genetics and Science and Its Times*. She is the author of *The Urinary System* in Greenwood's *Human Body Systems* series.

KELLI MILLER is an independent scholar and is the founder of NewScience, Inc. where she is feature writer on health and medicine for clients such as *The Scientist* and *Popular Science*.